BURROWING SHRIMPS AND SEAGRASS DYNAMICS IN SHALLOW-WATER MEADOWS OFF BOLINAO (NW PHILIPPINES)

Promotor: Prof. Dr. Wim van Vierssen
                          Professor of Aquatic Ecology
                          Wageningen University, The Netherlands

Co-promotor: Dr. Ir. Jan E. Vermaat
                          Associate Professor of Aquatic and Wetlands Ecology
                          Vrije Universiteit Amsterdam, The Netherlands

Awarding Committee: Prof. Dr. Han J. Lindeboom
                          Wageningen University, The Netherlands

                          Dr. Erik D. de Ruyter van Steveninck
                          UNESCO-IHE Institute for Water Education,
                          The Netherlands

                          Prof. Dr. Gerard van der Velde
                          Radboud University Nijmegen, The Netherlands

                          Prof. Dr. Wim J. Wolff
                          University of Groningen, The Netherlands

BURROWING SHRIMPS AND SEAGRASS DYNAMICS IN SHALLOW-WATER MEADOWS OFF BOLINAO (NW PHILIPPINES)

DISSERTATION
Submitted in fulfillment of the requirements of
the Academic Board of Wageningen University and
the Academic Board of UNESCO-IHE Institute for Water
Education for the Degree of DOCTOR
to be defended in public
on Friday, 26 September 2008 at 15:30 H in Delft,
The Netherlands

by

Hildie Maria Nacorda
born in Lucena City, Philippines

CRC Press/Balkema is an imprint of the Taylor & Francis Group, an informa business

Published by:
CRC Press/Balkema
PO Box 447, 2300 AK Leiden, The Netherlands
e-mail: Pub.NL@taylorandfrancis.com
www.crcpress.com – www.taylorandfrancis.co.uk – www.balkema.nl

ISBN 978-0-415-48402-2 (Taylor & Francis Group)
ISBN 978-90-8504-938-8 (Wageningen University)

Contents

Preface   vii

Summary   ix

Chapter 1   General introduction   1

Chapter 2   The distribution of burrowing shrimp disturbance in Philippine seagrass meadows
(to be submitted to Estuarine, Coastal and Shelf Science)   11

Chapter 3   Aboveground behavior of the snapping shrimp *Alpheus macellarius* Chace, 1988, and its significance for leaf and nutrient turnover in a Philippine seagrass meadow
(submitted to Journal of Experimental Marine Biology and Ecology)   31

Chapter 4   Burrows and behavior of the snapping shrimp *Alpheus macellarius*, Chace, 1988, in different seagrass substrates
(submitted to Marine Biology)   47

Chapter 5   Growth response of the dominant seagrass *Thalassia hemprichii* (Ehrenberg) Ascherson to experimental shrimp disturbance
(submitted to Marine Ecology Progress Series)   65

Chapter 6   General discussion and conclusions   83

Samenvatting   93

Buód (Summary in Filipino)   95

Kabuowan (Summary in Bolinao)   97

About the Author   99

Preface

This research project was carried out following the PhD sandwich scheme of the UNESCO-IHE Institute for Water Education in Delft with the Marine Science Institute (MSI) of the University of the Philippines in Diliman. The detailed proposal had been prepared and some laboratory analyses were carried out at IHE, the actual research has been implemented in the Philippines, mostly in Bolinao, Pangasinan, and all the chapters were completed and finalized at MSI in Diliman. A WOTRO fellowship through IHE (WB84-413) afforded me the opportunity to undertake this research project, which was supplemented later on by a scholarship contract from MERF-MSI, through the SARCS/ WOTRO/ LOICZ project *"Economic evaluation and biophysical modeling of the marine environment in Bolinao in support of management for sustainable use"*.

My sincere thanks to Dr. ir. Jan Vermaat, his extraordinary commitment, academic recommendations, constant guidance and motivation, and unfaltering support as adviser and co-promotor, and to Prof. Miguel Fortes, who, as co-adviser, gave *big brother* counsel and encouragement during various turning points. Jan and Sir Mike had endorsed my participation in the EU Project *CERDS* (*Responses of Coastal Ecosystems to Deforestation-derived Siltation in Southeast Asia*, TS3*–C T92–0301) and in various scientific meetings, all of which proved beneficial to the research project – the *5th International Crustacean Congress,* the *Seagrass Biology Workshops,* the *4th Symposium of the Marine Biology of the South China Sea,* and several symposia of the *Philippine Association of Marine Science* (PAMS).

Prof. Wim van Vierssen, my promotor, has provided the challenge to finish the thesis – thank you, Sir, for accepting the draft with a vote of confidence. I am grateful to Jan and Sir Mike who carefully read and critically reviewed all versions of the chapters in this book; the many refinements were from the advice of Dr. Erik de Ruyter van Steveninck, Prof. Dr. Wim J. Wolff, Prof. Dr. Gerld van der Velde, and Prof. Dr. Han Lindeboom. Some chapters also gathered comments from the experts – Dr. Johan Stapel, Dr. Eize Stamhuis, Dr. Charles Fransen, Prof. Annette Meñez, Dr. Rene Rollon, and Dr. Al Licuanan; Dr. Dosette Pante patiently explained the difficult details of multivariate statistics. Credit goes to Jan, Prof. Jojo Baquiran, and Marge Celeste, for providing the Dutch, Filipino, and Bolinao translations, respectively, of the English summary. At IHE, Erik de Ruyter, Erick de Jong, Jolanda Boots, and Laura Kwak took care of the administrative matters related to my promotion.

Dr. Laura David had set me up to finish – she pulled my tether string back to base, showed me how to start, kept me moving, and checked my writing/ re-writing progress everyday. Heartfelt thanks, Lau, and to my supervisors who were behind your mission – Profs. Malou McGlone, Gil Jacinto, and Cesar Villanoy; thanks Prof. Liana McManus for *Experiments in Ecology,* and, together with all senior staff at MSI who had loaned equipment and dropped their amusingly different versions of persuasion (they all worked!) – thank you for the reminder of not to lose sight of the goal during bottlenecks.

I am deeply indebted to all volunteers, field assistants, and colleagues who contributed time and effort during the field surveys and experiments in Bolinao, Palawan, and Naawan – Chris Ragos, Ronald de Guzman, Jack Rengel, Sheila Albasin, Napo Cayabyab, Andrea and Jonathan Persons, Dr. Sai Thampanya, Anjo Tiquio, Alvin Carlos, Day Lacap, Dr. Rev Alvarez-Molina, Jhun Castrence, Dr. Wili Uy, Freddie Lagarde, Lt. Jovy Bacordo, Karen Araño, Fe Pillos-Lomahan, Malou

Sison, Kat Villamor, Helen Bangi, Dave Pastor, Jaqui Pinat, Lani Verceles, Dr. Domeng Ochavillo, Angelo Pernetta, Ronald Gijlstra, Dr. José Vós, Nadia Palomar-Abesamis, Jovy Fomar, Ines Templo, Oytón Rubio Jr., Chà de Castro, Mímoy Silvano, Dr. Ruth Gamboa, Rex Montebon, Dr. Rommi Dizon, Makoy Ponce, Peter vander Wateren, and Frank Wiegman. Boyet Elefante[†] fabricated the stainless steel tripod for the underwater housing of the videocamera and designed the water supply of the aquaria used in the tank experiments while Cris Diolazo coordinated the fabrication of the sediment traps and the field 'mama' corer. Sheila also helped with data encoding while in Bolinao. At MSI in Diliman, Helen Dayao and Lot Luangco had helped augment and organize the literature collection (some of these have been kindly sent by Jan and Zayda Halun) and Arlene Boro attended to urgent needs from the University Registrar. Nicole van Beekum and Jean Brimbuela were very reliable with fellowship matters – thanks for keeping the transactions prompt.

My earnest gratitude also to Sai, for her friendship, and to the warm welcome to Europe by Mary Pernetta[†], Jan, Erik de Ruyter, Eize, Ineke Wesseling, Peter, and Marjolijne van Waveren-Hoerscht. I appreciated the constant company of fellow Filipino students then in Delft and Den Haag and of friends who visited – Ethel Balatayo, Paul Rivera, Rene, Karen, Day, Zayda, Dr. Hilly Ann Roa-Quiaoit, and Malou del Puerto-Villbrandt. Big thanks for the morále-boosting breaks by the Kahlúa Club, the Seagrass, Oceanography, and COMECO Labs at MSI, the magnanimous Aliño trio, Dosette, Rommi, Sam Mamauag, Catâ Rañola, and the badminton addicts. To Lau, Dosette, Dette de Venecia, Julie Otadoy, Badi Samaniego, and Rommi – thank you for the comforting words during the knock down spells. I also stumbled upon new friends – Alex Magallona, Chris Cromey, Jo Hernandez, Mon and Mely Romero, Caron de Mars, Michael Aw, Prof. Bill and Peggy Hamner, Gwen Noda, Lt. Jun Pretila, Mark Arboleda, and Joe Moreira – whose bold and sensible ideas that were all apt to relax the mind from its knots and warps.

Last but not least, a big hug to my family for the unreserved support (Raul and his family deserve special mention for the hospitality in Sydney), the incredible amount of patience (I apologize for having missed birthdays/ holidays), and inspiration through the years. I will always be grateful for my mother's advice and unceasing prayers; I was drawn to complete this work so that we can have stories to tell the children.

Hildie Maria E. Nacorda
January 2008

Summary

Natural disturbances contribute to the dynamics of seagrass meadows. This thesis intended to assess the importance of small-scale disturbance (bioturbation) by burrowing shrimps as a determinant of sediment mobility and as a factor limiting the establishment, expansion, and species composition of mixed-species beds in the Philippines. The specific aims of the studies were to describe the spatial distribution of burrowing shrimp disturbance in seagrass beds along a siltation gradient and its consequences to the vertical properties of bed sediments, to quantify the behavior of burrowing alpheid shrimps, and to determine the effects of short-term burial and leaf harvesting on the growth patterns of the dominant seagrass *Thalassia hemprichii* (Ehrenberg) Ascherson.

Burrow openings and sediment gaps were common in the meadows and were associated with either the caridean shrimp *Alpheus macellarius* Chace, 1988 or the larger, deeper burrowing species of Thalassinidea. Wave-sheltered beds harbored more frequent and larger sediment gaps than the exposed beds; shrimp burrows and mounds were also more prevalent in clear-water than in turbid seagrass areas. The disturbance by *A. macellarius* was found greater than that due to thalassinidean shrimps, overall, and alpheid distribution was found to be largely limited to vegetated fine-sand substrates. Areas of *A. macellarius*-reworked sand patches represented $15 \pm 2\%$ of the meadows (at scaled-up densities of 102 ± 5 per 100 m^2; cf. cover photograph) while sand mounds of thalassinidean shrimps covered $5 \pm 1\%$ (at densities of 52 ± 7 per 100 m^2). Burrowing shrimps altered the vertical profiles of sediment properties differently – *A. macellarius* relocated a significant proportion of coarse grains in the top 10 cm while thalassinidean shrimps consolidated finer fractions from 10 down to 20 cm of the sediments and concentrated organic matter. Both shrimps decreased sediment nitrogen by 20-73% in the top layer and by 4-46% at deeper than 10 cm. No significant changes in the sediments' phosphorus profiles were observed due to the shrimps.

In situ observations of the aboveground behavior of *A. macellarius* revealed its active sediment reworking and occasional harvesting of seagrass leaves during the dry months in a clear-water meadow. The shrimps allocated only 12% of its daily active period for these activities and were in their respective burrows during the remainder of daytime. In wet months, the rates of aboveground activities became reduced by at least 34% and within-burrow periods increased by 5% as a consequence. On average, *A. macellarius* remobilized ~300 g DW of sediment d^{-1} (or 112 kg y^{-1}), and harvested 0.8 g DW of leaves d^{-1} (or 291.3 g y^{-1}). The estimated sediment reworking rate is considerable (0.8 to 1.4 kg m^{-2} d^{-1}) for an average shrimp density of 2 individuals m^{-2}, and the leaf harvesting rate represents moderate herbivory (0.4 to 2.3 g m^{-2} d^{-1}), equivalent to 12 to 42% of leaf production.

Subsequent laboratory observations that examined the role of bed sediment type – sand, muddy sand, and sandy mud – on the behavior of *A. macellarius* showed that the shrimps immediately commenced with burrowing upon contact with the substrates. In the sand substrate, early success in concealment was achieved (first burrows within 2 hours) despite considerable burrowing effort and extensive tunnel lengths were also attained. Burrowing activities became reduced and wandering evident in all the substrates after the fifth week of observation. Feeding became conspicuous – mainly as particle ingestion, occasional suspension-feeding bouts, and direct grazing on seagrass leaves. Burrowing, grooming, and surveying behavior were predominant during the day while wandering and feeding were extended at night. Burrowing, however, remained significantly higher in sand than in the other two substrates. Overall, soft carbonate sand sediments together with reinforcement from dense seagrasses presented greater support for the burrowing behavior of *A. macellarius* and this is in line with the observed higher densities in these habitats in the field. In contrast, shrimp burrowing appeared limited and was substituted by concealment strategies in the terrigenous substrates, and, with less support from the sparse vegetation, burrow numbers were significantly lower in these habitats in the field.

The series of manipulative experiments mainly on the seagrass *T. hemprichii* provided evidence of the tolerance of vegetative shoots to the small-scale disturbance imposed by burrowing shrimps. In apical shoots, single burial events lasting at least 14 days induced accelerated leaf growth, while leaf clipping and combined treatments had minimal effects on either leaf or rhizome growth. Seedlings also survived defoliation but were sensitive to burial events – clipping alone did not cause changes in seedling growth but this significantly and continuously decreased with burial, applied alone and combined with clipping disturbance. Exclusion from shrimp activity did not influence leaf growth rates of mature shoots. Only *Halophila* densities were enhanced, particularly after 21 weeks of the 13-month lasting experiment, and both *T. hemprichii* growth and shoot densities of other coexisting seagrass species exhibited strong temporal variation as expected.

In short, this thesis established that the two types of burrowing shrimps – the alpheids and thalassinideans – that are common in the seagrass beds redistribute considerable quantities of sediment with significant substrate effects on depth gradients of organic matter, grain size, and nitrogen. Only the alpheids have a tight connection to the seagrass, i.e., they feed on it and remove a moderate component from primary production, but these do not affect the established clonal seagrass stands markedly. Probably they affect seed and seedling recruitment negatively, but this may not necessarily have a severe impact on the established stands these shrimps inhabit.

Chapter 1

General introduction

The ecosystem services provided by seagrass meadows, and valued at about a tenth of the total global flow value (Constanza et al. 1997), are central to current advocacies for resource restoration and conservation (Duarte 2002, Orth et al. 2006). Seagrass meadows are coastal ecosystems that often exhibit high primary production (Duarte and Chiscano 1999), which supports diverse floral and faunal assemblages and large marine animals (McConnaughey and McRoy 1979, Lewis and Stoner 1983, Howard et al. 1989, Hily and Bouteille 1999). Bed sediments and the leaf canopies serve as habitat and refugia to benthos and resident and transient fish (Bell and Westoby 1986, Bell and Pollard 1989, Connolly 1994, Loneragan et al. 1997, Sheridan 1997, Heck et al. 2003). Seagrass canopies modify currents and attenuate wave energy, a case of autogenic ecosystem engineering (*sensu* Jones et al. 1994), thus, trap sediment, seston, and larvae (Grizzle et al. 1996, Duarte et al. 1999, Koch 1999, Terrados and Duarte 2000, Vermaat et al. 2000, Gacia and Duarte 2001, Evrard et al. 2006), affect the storage of primary production within the system (Duarte and Cebrián 1996) or its export (Slim et al. 2006), and contribute to buffering adjacent sensitive habitats against the direct effects of water turbulence (Koch et al. 2006) and riverine siltation (Kenworthy et al. 1982, Cebrián et al. 2000). Belowground, the production of roots and rhizomes is substantial (45% of the meadow's total biomass; Vermaat et al. 1995), hence, probably fundamental to substrate stability (Duarte et al. 1998).

The persistence of seagrass meadows depends on vegetation processes and the plants' continuous response to various natural and anthropogenic disturbances. Seagrass meadows are dynamic systems composed of vegetative shoots in clonal networks that undergo demographic increases through branching, and decreases following senescence and mortality (Duarte et al. 2006). The recruitment of seeds and seedlings occurs periodically and may account for patch initiation (Duarte and Sand-Jensen 1990, Olesen et al. 2004). The rates of module production, ramet integration, colonization, and mortality vary with the size of the species (Marbá et al. 1996), habitat conditions, and season (Lee and Dunton 1996, Ramage and Schiel 1999). Disturbances may cause discontinuities in the landscape and affect these vegetation processes. These discontinuities are caused by medium- to large-scale natural events – dune migrations (Marbá et al. 1994a), waves and storms (Patriquin 1975, Fonseca and Bell 1988, Preen et al. 1995), and grazing (Ogden et al. 1973, Thayer et al. 1984, Valentine and Heck 1999, Sheppard et al. 2007) – that result in the dislodgment of the plants or their negative response to redistributions of sediments, eutrophication, and light reduction (Duarte et al. 1997, Terrados et al. 1998, Guidetti and Fabiano 2000).

Small-scale biological disturbance in seagrass meadows, prominent in benign environments, is caused by seagrass-associated animals (Jacobs et al. 1981, Hall et al. 1992, Philippart 1994, Valentine et al. 1994, Woods and Schiel 1997, Townsend and Fonseca 1998, Nacken and Reise 2000). Burrowing shrimps are a specific group of benthic invertebrates that may occur in association with seagrass environments, often conspicuous because of sediment gaps produced by their bioturbation (Fig. 1). In particular, the bioturbation by thalassinidean shrimps is

Figure 1. Evidence of burrowing shrimp disturbance in seagrass beds off Bolinao (NW Philippines) – sand mounds of thalassinidean shrimps (left photo) and sand patches of alpheid shrimps (right photo).

well-studied and known to affect plant dynamics (Suchanek 1983, Valentine et al. 1994, Duarte et al. 1997, Dumbauld and Willey-Echeverria 2006), patch expansion (Townsend and Fonseca 1998), the benthos (Branch and Pringle 1987, Berkenbusch et al. 2000, Pillay et al. 2007), and processes at the sediment-water interface (Koike and Mukai 1983, Murphy and Kremer 1992, Forster and Graf 1992, Forster 1996, Gilbert et al. 1998). Another group of burrowing shrimps, the alpheids (Caridea), could also dominate the range of biological interactions in seagrass ecosystems, e.g., *Alpheus edamensis* De Man in Barang Lompo (Indonesia), the activities of which were shown to prevent the export of organic matter, hence, contributed to the conservation of nutrients (Stapel 1997). These interactions are probably altered in mixed-species beds of tropical regions such as SE Asia, where human pressure on the coastal zone has increased due to development initiatives (Fortes 1988, Milliman and Syvitski 1992, Duarte 2002, Orth et al. 2006). The degree of sediment mobility determines the life span, extension, and horizontal mobility of seagrass patches (Marbá et al. 1994a, Vermaat et al. 1997), and the species that compose the beds differentiate along a size and age range that may be used to distinguish their response to siltation by clonal leaf and rhizome growth (Duarte et al. 1994, Vermaat et al. 1995). Sediment mobility is primarily driven by river inputs but at the lower end of the siltation gradient, biological modifications may become important, along with physical reworking caused by tidal currents, wind- and wave-induced resuspension, and bed load movement, all likely to be severe during monsoonal rains and storms.

The present thesis intends to examine the role of small-scale disturbance by burrowing shrimps and its interaction with seagrass performance against a changing background of anthropogenically altered sediment dynamics. It describes the results of fieldwork within the 27 km^2 seagrass beds off Santiago Island in Bolinao (northwestern Philippines) (McManus et al. 1995) and of outdoor experiments carried out in the hatchery facility of the Bolinao Marine Laboratory (BML) of the University of the Philippines-Marine Science Institute (UPMSI). These results of work conducted between June 1997 and April 2001 are reported in the succeeding four chapters. In Chapter 2, the distribution of burrowing shrimp disturbance (as apparent marks on the beds – burrows openings, pits, sand patches/ mounds – Fig. 1) in various seagrass and adjacent sediment environments is described. Apart from the beds on reef flats off the Bolinao-Anda region, the beds on selected islands of North Palawan, the Tubbataha Reef Atolls, and the reef flats off Naawan and Sulawan in northern Mindanao were visited to provide a comparative contrast to exposure conditions of the Bolinao-Anda region. The sediments disturbed by the shrimps are also characterized based on profiles of grain sizes, organic matter contents, and nutrients (nitrogen and phosphorus).

Chapter 3 quantifies the aboveground behavior of the snapping shrimp *Alpheus macellarius*, Chace, 1988 from its early morning emergence to prolonged retreat to the burrow by late afternoon to early evening. Observations were made *in situ* during three periods in a clear-water meadow off Silaki Island (Bolinao). The chapter also provides estimates of the significance of the shrimps' sediment reworking and seagrass harvesting. Because *A. macellarius* spent a considerable amount of time in its burrow based on field observations, an outdoor experiment was setup to determine the shrimp's behaviour within the burrow and to test the effect of sediment type on its activities. This study is organized as Chapter 4, which presents aspects of shrimp behaviour – burrow construction, feeding, and other within-burrow activities, activity pattern – as observed in glass cuvettes.

Chapter 5 contains results from a series of manipulative experiments that show the effects of short-term burial and leaf clipping on the growth patterns of the dominant seagrass *Thalassia hemprichii* (Ehrenberg) Ascherson. The manipulations involved monitoring leaf growth and rhizome elongation of apical shoots on clonal rhizome runners (Expt. 1, *in situ*) and leaf and root growth in seedlings (Expt. 2, outdoor). Experiment 3, also *in situ*, used exclosures to block all shrimp activity and tested the effect of the exclusion of shrimp disturbance on leaf growth and densities at the meter-scale. The thesis concludes with a synthesis (Chapter 6) that ties up results and conclusions in thematic sections, i.e., the role of bioturbation by burrowing shrimps in seagrass meadows, foraging strategies of *A. macellarius* and its mutualistic symbiosis with *Cryptocentrus* spp., shrimp disturbance and *T. hemprichii*, and, finally, small-scale disturbance and large-scale dynamics.

References

Bell, J.D. and Pollard, D.A. 1989. Ecology of fish assemblages and fisheries associated with seagrasses. In: Larkum, A.W.D., McComb, A.J., and Shepherd, S.A., eds., Biology of seagrasses, a treatise on the biology of seagrasses with special reference to the Australian region. Aquatic Plant Studies 2: 565-609. Elsevier, Amsterdam, The Netherlands.

Bell J.D. and Westoby M. 1986. Abundance of macrofauna in dense seagrass is due to habitat preference, not predation. Oecologia 68: 205-209.

Berkenbusch, J., Rowden, A.A., and Probert, P.K. 2000. Temporal and spatial variation in macrofauna community composition imposed by ghost shrimp *Callianassa filholi* bioturbation. Marine Ecology Progress Series 192: 249-257.

Branch, G.M., and Pringle, A. 1987. The impact of the sand prawn *Callianassa kraussi* Stebbing on sediment turnover and on bacteria, meiofauna and benthic microflora. Journal of Experimental Marine Biology and Ecology 107: 219-235.

Cebrián, J., Pedersen M.F., Kroeger, K.D., and Valiela, I. 2000. Fate of production of the seagrass *Cymodocea nodosa* in different stages of meadow formation. Marine Ecology Progress Series 204: 119-130.

Connolly, R.M. 1994. A comparison of fish assemblages from seagrass and unvegetated areas of a southern Australian estuary. Australian Journal of Marine and Freshwater Research 45: 1033-1044.

Constanza, R., d'Arge, R., Groot, R. de, Farber, S., Grasso, M., Hannon, B., Limburg, K., Naeem, S., O'Neill, R.V., Paruelo, J., Raskin, R.G., Sutton, P., and van den Belt, M. 1997. The value of the world's ecosystem services and natural capital. Nature 387: 253-260.

Duarte, C.M. 2002. The future of seagrass meadows. Environmental Conservation 29: 192-206.

Duarte, C.M. and Cebrián, J. 1996. The fate of marine autotrophic production. Limnology and Oceanography 41: 758-766.

Duarte, C.M. and Chiscano, C.L. 1999. Seagrass biomass and production: a reassessment. Aquatic Botany 65: 159-174.

Duarte, C.M. and Sand-Jensen, K. 1990. Seagrass colonization: patch formation and patch growth in *Cymodocea nodosa*. Marine Ecology Progress Series 65: 193-200.

Duarte, C.M., Benavent, E., and Sanchez, M.C. 1999. The microcosm of particles within seagrass (*Posidonia oceanica*) canopies. Marine Ecology Progress Series 181: 289-295.

Duarte, C.M., Fourqurean, J.W., Krause-Jensen, D., and Olesen, B. 2006. Dynamics of seagrass stability and change. In: Larkum, A.W.D., Orth, R.J., and Duarte, C.M., eds., Seagrasses: biology, ecology and conservation, pp. 271-294. Springer, Dordrecht, The Netherlands.

Duarte, C.M., Merino, M., Agawin, N.S.R., Uri, J., Fortes, M.D., Gallegos, M.E., Marbá, N., and Hemminga, M.A. 1998. Root production and belowground seagrass biomass. Marine Ecology Progress Series 171: 97-108.

Duarte, C.M., Terrados, J., Agawin, N.S.R., Fortes, M.D., Bach, S., and Kenworthy, W.J. 1997. Response of a mixed Philippine seagrass meadow to experimental burial. Marine Ecology Progress Series 147: 285-294.

Dumbauld B.R. and Wyllie-Echeverria, S. 2003. The influence of burrowing thalassinid shrimps on the distribution of intertidal seagrasses in Willapa Bay, Washington, USA. Aquatic Botany 77: 27-42.

Evrard V., Kiswara, W., Bouma, T.J., and Middleburg, J.J. 2005. Nutrient dynamics of seagrass ecosystems: [15]N evidence for the importance of particulate organic matter and root systems. Marine Ecology Progress Series 295: 49-55.

Fonseca, M.S. and Bell, S.S. 1988. Influence of physical setting on seagrass landscapes near Beaufort, North Carolina, USA. Marine Ecology Progress Series 171: 109-121.

Forster, S. 1996. Spatial and temporal distribution of oxidation events occurring below the sediment-water interface. P.S.Z.N. I: Marine Ecology 17: 309-319.

Forster, S. and Graf, G. 1992. Continuously measured changes in redox potential influenced by oxygen penetrating from burrows of *Callianassa subterranea*. Hydrobiologia 235/236: 527-532.

Fortes, M.D. 1988. Mangroves and seagrass beds of East Asia: habitats under stress. Ambio 17: 207-213.

Gilbert, F., Stora, G., and Bonin, P. 1998. Influence of bioturbation on denitrification activity in Mediterranean coastal sediments: an *in situ* experimental approach. Marine Ecology Progress Series 163: 99-107.

Grizzle, R.E., Short, F.T., Newell, C.R., Hoven, H., and Kindblom, L. 1996. Hydrodynamically induced synchronous waving of seagrasses: monami and its possible effects on larval mussel settlement. Journal of Experimental Marine Biology and Ecology 206: 165-177.

Guidetti, P. and Fabiano, M. 2000. The use of lepidochronology to assess the impact of terrigenous discharges on the primary leaf production of the Mediterranean seagrass *Posidonia oceanica*. Marine Pollution Bulletin 40: 449-453.

Hall, S.J., Raffaelli, D., and Thrush, S.F. 1992. Patchiness and disturbance in shallow water benthic assemblages. In: Giller, P.S., Hildrew, and A.G. Raffaelli, D.G., eds., Aquatic ecology: scale, pattern and process, Proceedings of the British Ecological Society and the American Society of Limnology and Oceanography Symposium, pp. 333-375. Blackwell Scientific Publications, Oxford.

Heck, K.L., Hays C.G., and Orth, R.J. 2003. A critical evaluation of the nursery role hypothesis for seagrass meadows. Marine Ecology Progress Series 253: 123-136.

Hily, C. and Bouteille, M. 1999. Modifications of the specific diversity and feeding guilds in an intertidal sediment colonized by an eelgrass meadow (*Zostera marina*) (Brittany, France). Life Sciences 322: 1121-1131.

Howard, R.K., Edgar, G.J., and Hutchings, P.A. 1989. Faunal assemblages of seagrass beds. In: Larkum, A.W.D., McComb, A.J., and Shepherd, S.A., eds., Biology of seagrasses, a treatise on the biology of seagrasses with special reference to the Australian region. Aquatic Plant Studies 2: 536-564. Elsevier, Amsterdam, The Netherlands.

Jacobs, R.P.W.M., den Hartog, C., Braster, B.F., and Carrier, F.C. 1981. Grazing of the seagrass *Zostera noltii* by birds at Terschelling (Dutch Wadden Sea). Aquatic Botany 10: 241-259.

Jones, C.G., Lawton, J.H., and Shachak, M. 1994. Organisms as ecosystem engineers. Oikos 69: 373-386.

Kenworthy, W.J., Zieman, J.C., and Thayer, G.W. 1982. Evidence for the influence of seagrasses on the benthic nitrogen cycle in a coastal plain estuary near Beaufort, North Carolina (USA). Oecologia 54: 152-158.

Koch, E.W. 1999. Sediment resuspension in a shallow *Thalassia testudinum* Banks *ex* König bed. Aquatic Botany 65: 269-280.

Koch, E.W., Ackerman, J.D., Verduin, J., and van Keulen, M. 2006. Chapter 8 – Fluid dynamics in seagrass ecology – from molecules to ecosystems. In: Larkum, A.W.D., Orth, R.J., and Duarte, C.M., eds., Seagrasses: biology, ecology and conservation, pp. 193-225. Springer, Dordrecht.

Koike, I. and Mukai, H. 1983. Oxygen and inorganic contents and fluxes in burrows of the shrimps *Callianassa japonica* and *Upogebia major*. Marine Ecology Progress Series 12: pp. 185-190.

Lee. K.-S., and Dunton, K.H. 1996. Production and carbon reserve dynamics of the seagrass *Thalassia testudinum* in Corpus Christi Bay, Texas, USA. Marine Ecology Progress Series 143: 201-210.

Lewis, F.G. III and Stoner, A.W. 1983. Distribution of macrofauna within seagrass beds: an explanation for patterns of abundance. Bulletin of Marine Science 33: 296-304.

Loneragan, N.R., Bunn, S.E., and Kellaway, D.M. 1997. Are mangroves and seagrasses sources of organic carbon for penaeid prawns in a tropical Australian estuary? A multiple stable isotope study. Marine Biology 130: 289-300.

Marbá N., Cebrián, J., Enriquez, S., and Duarte, C.M. 1994a. Migration of large-scale subaqueous bedforms measured with seagrasses (*Cymodocea nodosa*) as tracers. Limnology and Oceanography 39: 126-133.

Marbá, N., Cebrián, J., Enríquez, S., and Duarte, C.M. 1996. Growth patterns of Western Medditeranean seagrasses: species-specific responses to seasonal forcing. Marine Ecology Progress Series 133: 203-215.

Marbá, N., Gallegos, M.E., Merino M., and Duarte, C.M. 1994b. Vertical growth of *Thalassia testudinum*: seasonal and interannual variability. Aquatic Botany 47: 1-11.

McConnaughey, T. and McRoy, C.P. 1979. [13]C label identifies eelgrass (*Zostera marina*) carbon in an Alaskan estuarine food web. Marine Biology 53: 263-269.

McManus, J.W., Nañola, C.L., Reyes, R.B. Jr., and Kesner, K.N. 1995. The Bolinao coral reef resource system. In: Juinio-Meñez, M.A. and Newkirk G.F., eds., Philippine coastal resources under stress, selected papers from the 4th Annual Common Property Conference, Manila, Philippines, 16-19 June 1993, pp. 193-204.

Milliman, J.D. and Syvitski, J.P.M. 1992. Geomorphic/ geotectonic control of sediment discharge to the ocean – the importance of small mountainous rivers. Journal of Geology 100: 525-544.

Murphy, R.C. and Kremer, J.N. 1992. Benthic community metabolism and the role of deposit-feeding callianassid shrimp. Journal of Marine Research 50: 321-340.

Nacken, M. and Reise, K. 2000. Effects of herbivorous birds on intertidal seagrass beds in the northern Wadden Sea. Helgoland Marine Research 54: 87-94.

Ogden, J.C., Brown, R.A., and Salesky, N. 1973. Grazing by the echinoid *Diadema antillarum* Philippi: formation of halos around West Indian patch reefs. Science 182: 715-717.

Olesen, B., Marbá, N., Duarte, C.M. Savela, R.S., and Fortes, M.D. 2004. Recolonization dynamics in a mixed seagrass meadow: the role of clonal versus sexual processes. Estuaries 27: 770–780.

Orth, R.J., Caruthers, T.J.B., Dennison, W.C., Duarte, C.M., Fourqurean, J.W. , Heck, K.L. Jr., Hughes, A.R., Kendrick, G.A., Kenworthy, W.J., Olyarnik, S., Short, F.T., Waycott, M., and Williams, S.L. 2006. A global crisis for seagrass ecosystems. BioScience 56: 987-996.

Patriquin, D.G. 1975. "Migration" of blowouts in seagrass beds at Barbados and Carriacou, West Indies, and its ecological and geological implications. Aquatic Botany 1: 163-189.

Philippart, C.J.M. 1994. Interactions between *Arenicola marina* and *Zostera noltii* on a tidal flat in the Dutch Wadden Sea. Marine Ecology Progress Series 111: 251-257.

Pillay, D., Branch, G.M., and Forbes, A.T. 2007. The influence of bioturbation by the sandprawn *Callianassa kraussi* on feeding and survival of the bivalve *Eumarcia paupercula* and the gastropod *Nassarius kraussianus*. Journal of Experimental Marine Biology and Ecology 344: 1-9.

Preen, A.R., Lee Long, W.J., and Coles, R.G. 1995. Flood and cyclone related loss, and partial recovery of more than 1000 km^2 of seagrass in Hervey Bay, Queensland, Australia. Aquatic Botany 52: 3-17.

Ramage, D.L. and Schiel, D.R. 1999. Patch dynamics and response to disturbance of the seagrass *Zostera novazelandica* on intertidal platforms in southern New Zealand. Marine Ecology Progress Series 189: 275-288.

Sheppard, J.K., Lawler, I.R., and Marsh, H. 2007. Seagrass as pasture for seacows: landscape-level dugong habitat evaluation. Estuarine, Coastal and Shelf Science 71: 117-132.

Sheridan, P. 1997. Benthos of adjacent mangrove, seagrass and non-vegetated habitats in Rokkery Bay, Florida, USA. Estuarine, Coastal and Shelf Science 44: 455-469.

Slim, F.J., Hemminga, M.A., Cocheret De La Morinière, E., and Van Der Velde, G. 1996. Tidal exchange of macrolitter between a mangrove forest and adjacent seagrass beds (Gazi Bay, Kenya). Aquatic Ecology 30: 119-128.

Stapel, J. 1997. Nutrient dynamics in Indonesian seagrass beds: factors determining conservation and loss of nitrogen and phosphorus, pp. 33-41. PhD thesis, Katholieke Universiteit Nijmegen. WOTRO/ NWO, 127 p.

Suchanek, T.H. 1983. Control of seagrass communities and sediment distribution by *Callianassa* (Crustacea, Thalassinidea) bioturbation. Journal of Marine Research 41: 281-298.

Terrados, J. and Duarte, C.M. 2000. Experimental evidence of reduced particle resuspension within a seagrass (*Posidonia oceanica* L.) meadow. Journal of Experimental Marine Biology and Ecology 243: 45-53.

Terrados, J., Duarte, C.M., Fortes, M.D., Borum, J., Agawin, N.S.R., Bach, S., Thampanya, U., Kamp-Nielsen, L., Kenworthy, W.J., Geertz-Hansen, O., and Vermaat, J. 1998. Changes in community structure and biomass along gradients of siltation in SE Asia. Estuarine, Coastal and Shelf Science 46: 757-768.

Thayer, G.W., Bjorndal, K.A., Ogden, J.C., Williams, S.L., and Zieman, J.C. 1984. Role of larger herbivores in seagrass communities. Estuaries 7: 351-376.

Townsend E.C. and Fonseca, M.S. 1998. Bioturbation as a potential mechanisms influencing spatial heterogeneity of North Carolina seagrass beds. Marine Ecology Progress Series 169: 123-132.

Valentine, J.F. and Heck, K.L. Jr. 1999. Seagrass herbivory: evidence for the continuous grazing of marine grasses. Marine Ecology Progress Series 176: 291-302.

Valentine, J.F., Heck, K.L., Harper, P., and Beck, M. 1994. Effects of bioturbation in controlling turtlegrass (*Thalassia testudinum* Banks *ex* König) abundance: evidence from field enclosures and observations in the Northern Gulf of Mexico. Journal of Experimental Marine Biology and Ecology 178: 181-192.

Vermaat, J.E., Agawin, N.S.R., Duarte, C.M., Enríquez, S., Fortes, M.D., Marbá, N., Uri, J.S., and van Vierssen, W. 1997. The capacity of seagrasses to survive increased turbidity and siltation: the significance of growth form and light use. Ambio 26: 499-504

Vermaat, J.E., Agawin, N.S.R., Duarte, C.M., Fortes, M.D., Marbá, N., and Uri, J.S. 1995. Meadow maintenance, growth and productivity of a mixed Philippine seagrass bed. Marine Ecology Progress Series 124: 215-225.

Vermaat, J.E., Santamaria, L., and Roos, P.J. 2000. Water flow across and sediment trapping in submerged macrophyte beds of contrasting growth form. Archiv für Hydrobiologie 148: 549-562.

Woods, C.M.C. and Schiel, D.R. 1997. Use of seagrass *Zostera novazelandica* (Setchell, 1933) as habitat and food by the crab *Macrophthalmus hirtipes* (Heller, 1862) (Brachyura, Ocypodidae) on rocky intertidal platforms in southern New Zealand. Journal of Experimental Marine Biology and Ecology 214: 49-65.

Chapter 2

The distribution of burrowing shrimp disturbance in Philippine seagrass meadows

H.M.E. Nacorda[1,2], N.M. Cayabyab[1], M.D. Fortes[1], J.E. Vermaat[2,3]

[1] Marine Science Institute, University of the Philippines, UPPO Box 1, Diliman, Quezon City, 1101, The Philippines
[2] Department of Environmental Resources, UNESCO-IHE Institute for Water Education, PO Box 3015, 2601 DA Delft, The Netherlands
[3] present address: Institute for Environmental Studies, Vrije Universiteit, De Boelelaan 1087, 1081 HV Amsterdam, The Netherlands

Abstract

Small-scale disturbance of seagrass meadows by burrowing shrimps was assessed by mapping and quantifying apparent disturbance marks (burrow openings, sand patches, sand mounds, shafts) and obtaining vertical profiles of sediment properties (grain size composition, organic matter and nutrient contents). The densities and sizes of sediment gaps and burrow openings were determined within sampling quadrats in various meadows and were correlated with ambient bed and site characteristics. Effects on sediment properties were determined in a comparison of disturbed and undisturbed areas in two beds with contrasting organic matter sources. Burrow openings and sediment gaps were common in all the meadows and were associated with either the snapping shrimp *Alpheus macellarius* Chace, 1988 (Alpheidae, Caridea), or species of Thalassinidea. Alpheid shrimp disturbance was more frequent than that of thalassinidean shrimps. The distribution of sand mounds and associated shafts of thalassinidean shrimps were random, whereas the distribution of sand patches and burrow openings of alpheid shrimps on the beds appeared regular and clumped, respectively. The densities and sizes of sediment gaps and openings were higher and larger in wave-protected than in exposed beds ($p < 0.05$). Shrimp disturbance was more prevalent in clear-water than in turbid seagrass areas of the Bolinao-Anda region. The distribution of thalassinidean shrimps, however, was wider than that of *A. macellarius*, which was observed to be limited to vegetated fine-sand substrates. Overall, alpheid shrimps reworked sand patch areas of $14 \pm 2\%$ of the meadows while thalassinidean shrimps produced sand mounds that covered $4 \pm 1\%$.

Burrowing shrimps altered the vertical profiles of sediment properties – *A. macellarius* relocated a significant proportion of coarse grains in the upper 10 cm while the thalassinidean shrimps consolidated finer fractions from 10 to 20 cm down the core, and consequently concentrated organic matter. Both shrimps reduced nitrogen by 20-73% in the top 10 cm sediment layer, by 4-36% in sediment deeper than 10 cm, but did not affect profiles of phosphorus.

Keywords: sediment gaps, *Alpheus macellarius*, Thalassinidea, burrows, seagrass canopies, sediment characteristics, Philippines

Introduction

Discontinuities in seagrass landscapes can result from disturbances that can have natural causes, such as rapid currents and exposure to waves (Patriquin 1975, Fonseca et al. 1983, Fonseca and Bell 1988, Hemminga and Duarte 2000), cyclones (Preen et al. 1995), as well as grazing by large mammals (De Iong et al. 1995, Preen 1995, Sheppard et al. 2007), rays (Valentine et al. 1994, Townsend and Fonseca 1998), and urchins (Ogden et al. 1973, Heck and Valentine 1995, Rose et al. 1999). Man-made causes of discontinuities include blast fishing (Jennings and Kaiser 1998), boat moorings (Walker et al. 1989, Hastings et al. 1995, Francour et al. 1999), and propeller scars (Dawes et al. 1997, Kenworthy et al. 2000). Small discontinuities in the meadows are due to disturbance by lugworms (Philippart 1994), crabs (Woods and Schiel 1997, Townsend and Fonseca 1998), and herbivorous birds (Jacobs et al. 1981, Nacken and Reise 2000), but sediment gaps – sand patches and mounds – result primarily from bioturbation by burrowing shrimps (Suchanek 1983, Stapel et al. 1997, Dumbauld and Wyllie-Echeverria 2003). The mechanisms by which small-scale shrimp disturbance contributed to the maintenance of one tropical clear-water seagrass meadow were determined by Duarte et al. (1997) in their assessment of each species' response – shoot density, vertical growth, branching – to sediment burial. However, the more general extent of such disturbance has remained unclear in coastal systems such as in the Southeast Asian region, where siltation events often threaten coastal habitats (Milliman and Meade 1983, van Katwijk et al. 1993, McClanahan and Obura 1997, Terrados et al. 1998, Barnes and Lough 1999, Wesseling et al. 2001).

Burrowing shrimps generate small- to intermediate-scale disturbance (Hall et al. 1992). They modify the sediment environment primarily through their foraging and turbative behavior. Most deposit-feeding thalassinidean shrimps (Thalassinidea, Decapoda) produce conspicuous mounds and construct shallow cylindrical shafts and funnels around the mounds (Nickell and Atkinson 1995). The mounds arise from the shrimps' ejections of fine sediments from belowground (Waslenchuk et al. 1983, Stamhuis et al. 1996), a behavior that has been associated with food sorting (De Vaugelas and Buscail 1990, Nickell and Atkinson 1995) and burrow ventilation (Stamhuis et al. 1996). The shafts are sites where the shrimps obtain food from surface sands while simultaneously maneuvering freely (De Vaugelas and Buscail 1990). Some species of alpheid shrimps (Alpheidae, Caridea, Decapoda) dump tunnel sediments on the bed surface through the burrow openings and stack rubble and shell fragments opposite to the dumping site (Karplus 1987). The dumping of sediments is a typical penultimate behavioral state during tunnel extension and appears as a means by which alpheid shrimps access the sediment surface. The stack of rubble and shells, which resembles a roof and which may continue down as the burrow wall, provides structural support to the burrow opening (Karplus 1987, Dworschak and Ott 1993).

We hypothesized that where high densities of burrowing shrimps occur, sediment properties and processes may be affected at the larger meadow scale (e.g., Chapin et al. 1997). As an initial approach to address this issue, therefore, we implemented a survey programme that assessed the presence of apparent disturbance marks in various seagrass environments. We related the distributions to bed characteristics to serve as a qualitative basis for the estimation of the significance of these animals for sediment disturbance and seagrass herbivory in our subsequent work (Nacorda et al., Chapter 3, this Thesis). We also looked at profiles of sediment attributes to describe the shrimps' disturbance region and the extent by which the animals particularly altered distributions of grain sizes and organic matter as well as concentrations of nitrogen and phosphorus in the sediments.

Materials and Methods

Study sites
We surveyed selected shallow seagrass areas in three locations – north-western, western, and southern Philippines (Fig. 1, a-c). These beds were on wave-protected reef flats (i.e., Bolinao-Anda and northern Mindanao), leeward zones of islands (Nangalao) or open to wind and waves (Pangaldaoan, northern Palawan and in the Tubbataha Reef atolls) (Table 1). Within the Bolinao-Anda region (I), we also examined unvegetated substrates of lagoons in the reef flat (Sites 3 and 5), which were adjacent to the seagrass sites, and of sandy areas that followed reef slopes (Sites 1, 6, and 14) (Table 1).

Mapping of apparent shrimp disturbance and calculations
Rope quadrats (n = 3, size 9 to 25 m^2, laid 5 m apart and perpendicular to the shore) were set and corners were pegged into the substrate, then 1-m^2 grids were made within each quadrat to facilitate mapping. On rapid surveys in the Tubbataha Reefs and in Lopez Jaena (Misamis Occidental), short transects of 50-m and 20-m lengths, respectively, were installed instead, and mapping involved a 1-m^2 steel quadrat laid at 5-m intervals on both sides of the transect.

Systematic mapping proceeded by plotting the positions of thalassinidean shrimp sand mounds, funnels and shafts, alpheid shrimp burrows, and surrounding sand patches in all the 1-m^2 squares within the sampling quadrat. Measures of the following were determined: base diameter for sand mounds, longest length (L), and widest width (W) for sand patches, and diameters for burrows and shafts. Based on these measurements, the area was calculated, which was used to indicate the potential effects to the meadows in question. The base area of sand mounds covered a circle, hence:

$$A = \Pi r^2,$$

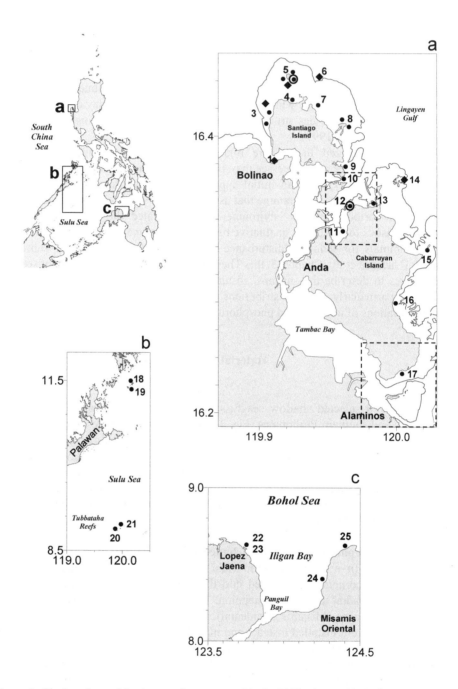

Figure 1. The locations of the three regions surveyed in the Philippines – (a), Bolinao-
    Anda reef system (Sites 1, 3-17); (b), North Palawan islands (Sites 18-19) and the
    Tubbataha Reef atolls (Sites 20-21); and (c), Iligan Bay in northern Mindanao (Sites 22-
    25). The specific seagrass habitats surveyed are indicated by (●) and the unvegetated
    habitats by (◆). Boxed areas in map (a) have turbid water; sediment core samples for
    detailed characterization were collected from two seagrass sites marked (◉, 5 and 12).
    (Base maps courtesy of R. Abesamis and P. van der Wateren)

Table 1. Broad site characteristics of areas examined for the distribution of burrowing shrimp disturbance.

Region/ Sites	Site description
I. Bolinao-Anda seagrass (● in Fig. 1, a)	Waters from the South China Sea and the Lingayen Gulf overlying the seagrass-dominated reef flat that is protected from outside waves by an intertidal reef crest (McManus et al. 1992); siltation gradient apparent from clear-water Site 5 (Silaki Is.; 16.4435 N, 119.9214 E) to turbid Site 17 (Batiarao; 16.2282 N, 120.0042 E) (McManus and Chua 1990); turbid conditions in Sites 9 (Pislatan; 16.3781 N, 119.9626 E) to 13 (Carot; 16.3496 N, 119.9650 E) also due to the organically loaded water mass of the Caquiputan Channel (Rivera 1997) and to discharges from fishponds and from the Sta. Rita River in Cabarruyan Island (Anda); bed substrate 90% coarse and fine sand with at least 8% silt and 4% organic matter (Kamp-Nielsen et al. 2002); mixed seagrass community dominated by *Thalassia hemprichii* (Ehrenberg) Ascherson (Vermaat et al. 1995) and declined in species richness above a 15% silt threshold (Terrados et al. 1998); sites intertidal to 5 m deep
I. Bolinao-Anda unvegetated substrata (♦ in Fig. 1, a)	Site 1 (Guiguiwanen; 16.3823 N, 119.9112 E) within an embayment, clear-water to turbid, with freshwater input from groundwater; Site 3 (Lucero; 16.4235 N, 119.9044 E) on western reef flat of Santiago Island, clear-water; both Sites 1 and 3 with muddy-sandy substrates and depths of 3 to 5 m; Site 6 (Malilnep; 16.4430 N, 119.9439 E) within a channel and Site 14 (Cangaluyan; 16.3687 N, 120.0059 E) after the reef slope
II. North Palawan Islands seagrass (● in Fig. 1, b)	Overlying water from the Sulu Sea, clear; Site 18 (Pangaldaoan; 11.4931 N, 120.1426 E) exposed to waves, and with short *T. hemprichii* and *Cymodocea rotundata* Ehrenberg *et* Hemprich *ex* Ascherson on coarse coralline substrata; Site 19 (Nangalao; 11.3478 N, 120.1633 E) at the leeward side of the island with stands of mostly *Enhalus acoroides* (L.f.) Royle and *T. hemprichii* on white sandy substrate mixed with coral rubble
II. Tubbataha Reef Atolls seagrass (● in Fig. 1, b)	Sulu Sea clear water also overlying the seagrass-dominated reef flats; both Sites 20 (South Islet; 8.8756 N, 119.8730 E) and 21 (North Islet; 8.9616 N, 119.9767 E) with white sandy substrates mixed with coral rubble; short shoots of *T. hemprichii* and *C. rotundata* found
III. Northern Mindanao seagrass (● in Fig. 1, c)	Bohol Sea water overlying the seagrass-dominated reef flats; Sites 22 (Kapayas Island) and 23 (Danlugan) in Lopez Jaena (8.6206 N, 123.7527 E) both on the west coast of Iligan Bay; Site 22 more seaward, exposed, and with clearer water than Site 23; *T. hemprichii* dominant in Site 22, *E. acoroides* in Site 23; Sites 24 (Naawan; 8.4061 N, 124.2620 E) and 25 (Sulawan; 8.6244 N, 124.3999 E) flanked the outer east coast of Iligan Bay (Misamis Oriental); Site 24 nearshore and on a narrow reef flat, with water turning turbid due to river discharge; dense population of *Halodule uninervis* (Forsskal) Ascherson found, occasionally with *C. rotundata* and *Syringodium isoetifolium* (Ascherson) Dandy (Uy et al. 2001); Site 25 in a remote cove of Tabajon (Lagundingin), a clear-water area with ~110 ha of mixed seagrasses dominated by *T. hemprichii* (Arriesgado 1999)

while the area of irregularly shaped sand patches was estimated from

$$A = L \times W,$$

an approach applied to similarly shaped corals (English et al. 1994).

Seagrass species richness and densities for each site were determined from either samples collected using a Rambo corer (diameter = 20 cm, area = 0.03 m^2; n = 10), *in situ* counts within 50 x 50 cm quadrats (n = 3; haphazardly thrown outside the mapped rope quadrat), or obtained from literature, in the case of sites in northern Mindanao, i.e., Arriesgado (1999) and Uy et al. (2001). Sediment samples were also collected for the determination of grain size structure (corer diameter 4 cm, length = 10 cm; n = 3); water depths at sampling time were recorded in all sites.

Core sampling, sediment characterization, and laboratory analyses

A separate set of sediment samples were obtained from Sites 5 (Silaki; clear-water; K_d from 0.1 to 0.7 m^{-1}, Rollon 1998) and 12 (Rufina; turbid; K_d from 0.1 to 2.0 m^{-1}, Rollon 1998) (Fig. 1, a) using PVC corers (10 cm diameter x 35 cm length). The samples were collected near sand patches with burrow openings (core type #1, n = 3), sand mounds with shafts (core type #2, n = 3), and on the homogeneously vegetated section of the meadow (core type #3, n = 3). All the core samples were kept upright and soaked in ambient water during transport to the laboratory.

Each core sample was sliced into sediment sections at depths 3, 6, 9, 12, 15, 20, 25, 30, and 35 cm from the surface. Each section was rid off of plant matter, sub-sampled for total organic matter (TOM) and nutrient analyses, then oven-dried altogether at 105°C for 24 h, cooled, and weighed. TOM was determined as material lost from the dried sample after ignition for 6 h at 550°C (Buchanan 1984). The rest of the each slice was analyzed for grain size distribution, i.e., characterized by wet sieving after soaking up to 100 g of dried samples in Calgon (sodium hexametaphosphate; Buchanan 1984). Fractions retained on each sieve were similarly oven-dried and weighed as described above. The grain size profile for each site was summarised as mean (μ, ϕ), standard deviation or sorting (σ, ϕ), and skewness (Leeder 1982).

Nitrogen (N) and phosphorus (P) of sediments were analyzed at the Environmental Engineering Laboratory of UNESCO-IHE. Known weights of oven-dried sediment samples (grain size <1 mm) were first digested with a mixture of sulphuric acid, selenium, and salicylic acid (Kruis 2000) prior to the analyses. Total N (as NH_4-N) was determined from the resulting digests following straightforward procedures and utilized a Perkin-Elmer Lambda 20/2.0 nm UV/VIS spectrometer. The standard addition technique was utilized for the analyses of total P (as PO_4-P) in a Tecator-Aquatec autoanalyser (APHO 1992 in Kruis 2000).

Data analyses

Each apparent disturbance point (burrow openings, sand patches, sand mounds, shafts) on seagrass areas of Region I was scored and analyzed for pattern using quadrat analysis, where variance-mean ratios (VMR) are calculated then interpreted, i.e., VMR approximates 1 for random distributions, <1 for uniform, and >1 for clustered distributions (Rogerson 2001). Mixed analyses of variance in SPSS were used to test the quadrat dataset of all sites. Densities of the disturbance marks were compared among regions (I-III), among habitat types (seagrass, reef flat lagoon, reef lagoon), and among sites for the Bolinao-Anda region (I). Sampling depth, sediment descriptors, mud content, seagrass parameters, and sediment gap densities and sizes were entered as covariates during the analyses and their sums of squares remained additive. *Post-hoc* comparisons were carried out where regional or habitat differences emerged, maintaining an experimentwise error rate of 0.017 (Sokal and Rohlf 1995). Appropriate transformations were carried out on heteroscedastic data.

Attributes in the sediment dataset were compared across core types and sections for each site using two-way ANOVAs. *Post-hoc* comparisons were similarly carried out as described.

Results

Mapping survey – comparisons of regions

The shallow meadows surveyed had sandy substratum, were poorly sorted, and were fine-skewed except for Region II which was coarse-skewed (Table 2). Substrate type was finer in the deeper and unvegetated lagoons of the Bolinao-Anda region (I), indicating low energy environments comparable to the meadows surveyed. The mixed meadows visited were dominated by *Thalassia hemprichii*, and those composed of as much as 7 species occurred in the relatively protected meadows (Table 2).

The marks of shrimp disturbance were evident in all the sites surveyed (Table 3). Some sand mounds occurred within huge sand patches and some alpheid shrimp burrows opened from mound margins as well. The burrows within the sand patches were inhabited by *Alpheus macellarius,* Chace, 1988, which live in symbiosis with species of *Cryptocentrus.* We noted the presence of craters of fish (e.g., *Amblygobius phalaena*) and crabs (e.g., *Callinectes* sp.) in silty environments, and of pits occupied by *Corallianassa* sp. in the intertidal zone of Site 3 (Lucero) (Fig. 1, a) and in wave-exposed Site 18 (Pangaldaoan) (Fig. 1, b).

The exposed sites of Region II had reduced densities and sizes of sand patches, burrow openings, and mound-associated shafts (Table 3). Sand mounds were low and appeared as 'moon-scapes' that covered $9 \pm 4\%$ of the bed. In the protected beds of Regions I and III, we observed our sampling quadrat of 9 m^2 to include 9 ± 1 sand patches of alpheid shrimps (maximum = 2 m^{-2}) with altogether 20 ± 5 burrow openings (maximum = 10 m^{-2}), 3 ± 1 sand mounds of thalassinidean shrimps (maximum = 3 m^{-2}), and 3 ± 1 mound-associated shafts (maximum = 3 m^{-2}) within funnel-shaped surface openings. The sand patches bared $13 \pm 2\%$ of the bed (maximum = 52%) and its burrows opened $\sim 0.1 \pm 0.02\%$ of the patch surface (maximum = 1%). Mean sand mound height was 9 ± 0.3 cm from the bed (maximum = 48 cm). Sand mound cover was significantly lower at $4 \pm 0.7\%$ of the meadows (maximum = 32%) compared with mound cover in Region II. Mound-associated shafts opened $0.05 \pm 0.01\%$ of the bed surface (maximum = 0.5%).

Sand patch numbers and sizes positively correlated with the number of burrow openings (Fig. 2, a and b). Mound sizes covaried positively with corresponding densities and with water depth (Table 4). There were also significant among-site variations in seagrass species richness and shoot densities of *T. hemprichii, H. uninervis*, and *E. acoroides* ($p < 0.05$, Table 4). Sediment skewness emerged as a positive correlate of seagrass species richness and the presence of sand mounds seemed to indicate a negative effect on shoot densities of *T. hemprichii* but

Table 2. Mean values of ($\pm$ SEM) depth, sediment characteristics, species richness and shoot densities of seagrasses on the 3 regions (25 sites) surveyed in the Philippines. Legend: (a), the species was found in the meadow but was missed in the quadrats or cores; –, the species was not found in the quadrats or cores; superscripts denote similar or significant differences among means after *post-hoc* comparisons at $\alpha = 0.017$)

REGION	I			II	III
Parameters	Bolinao-Anda reef system (NW Phils)			N Palawan & Tubbataha Reefs	N Mindanao (S Phils)
	After reef slope	Reef flat lagoon	Seagrass	Seagrass	Seagrass
Sampling depth, m	4.65 (0.26)	4.12 (0.17)	0.89 (0.06)	0.84 (0.22)	0.69 (0.14)
Sediment parameters:					
Mean grain size, μ, in ϕ	+ 2.39[b] (0.32)	+ 1.48[b] (0.02)	+ 1.17[a] (0.06)	0.69 (0.07)	+ 3.24 (0.56)
Sorting, σ, in ϕ	+ 1.81[a] (0.22)	+ 2.35[b] (0.36)	+ 1.92[a] (0.06)	2.44 (0.11)	+ 3.75 (0.48)
Skewness	+ 0.21 (0.06)	+ 0.72 (0.01)	+ 0.06 (0.02)	– 0.13 (0.01)	+ 0.48 (0.04)
Mud content, %	10.4 (2.4)	9.9 (2.2)	5.2 (0.4)	5.9 (0.5)	34.6 (7.5)
Seagrass parameters:					
Number of species	no seagrass	no seagrass	4 (0.01)	2 (0.1)	4 (0.6)
Shoot densities, per sq. m					
Cymodocea rotundata			36 (6)	88 (13)	33 (19)
Cymodocea serrulata			27 (4)	–	32 (22)
Enhalus acoroides			22 (2)	(a)	57 (19)
Halophila ovalis			15 (4)	19 (17)	20 (8)
Halodule uninervis			32 (6)	–	466 (287)
Syringodium isoetifolium			27 (6)		210 (133)
Thalassia hemprichii			180[a] (18)	622[b] (123)	516[b] (138)

Table 3. Mean density and space occupied (± SEM) by sediment gaps of burrowing shrimps in the three regions (25 sites) surveyed and in 3 habitats of the Bolinao-Anda region. Differences among regions were detected by ANOVA for alpheid shrimp sand patch densities and subsequent sizes, as well as sizes of thalassinidean shrimp sand mounds. Sand mound densities and sizes of alpheid shrimp burrow openings, sand mounds, and shaft opening sizes also differed across substrata in Region 1. Results of subsequent *post hoc* comparisons of regions are indicated by the lowercase superscripts, and of substrate comparisons by uppercase superscripts ($\alpha = 0.017$).

REGION	I			II	III
VARIABLES	Bolinao-Anda reef system (NW Phils)			N Palawan Islands & Tubbataha Reefs	N Mindanao (S Phils)
	After reef slope	Reef flat lagoon	Seagrass	Seagrass	Seagrass
Number of quadrats sampled	16	8	71	8	10
Densities (numbers/ 9 m^2)					
Alpheid shrimp sand patches VMR$^+$ (range)	-	-	9^b (0.5) 0.1 – 0.7	0.5^a (0.3)	7^b (2)
Alpheid shrimp burrow openings* VMR	31 (9)	33 (21)	2^b (2) 1.3 – 4.8	7^a (4)	20^b (9)
Thalassinidean shrimp sand mounds VMR (range$^+$)	29^B (8)	19^B (3)	4^A (0.5) 0.8 – 1.2	3 (1)	2 (1)
Mound-associated shafts VMR	6 (1)	5 (2)	3^b (0.6) 0.8 – 1.2	0.2^a (0.1)	4^b (1)
Sizes (% of space occupied/ 9 m^2)					
Alpheid shrimp sand patches	-	-	15.8^b (1.8)	0.4^a (0.3)	10.0^b (3.0)
Alpheid shrimp burrow openings	0.7^B (0.4)	1.0^B (0.9)	0.1^A (0.03)	0.09 (0.02)	0.08 (0.02)
Thalassinidean shrimp sand mounds	10.6^B (2.9)	16.6^B (5.1)	3.3$^{A;a}$ (0.6)	9.2^b (3.6)	2.5^a (0.9)
Mound-associated shafts/ pits	0.1^B (0.1)	0.2^B (0.00)	0.03^A (0.01)	0.001 (0.001)	0.08 (0.04)

Legend: $^+$VMR – variance to mean ratio, here given as the range of the most frequent pattern; * *Alpheus macellarius* – *Cryptocentrus* sp. burrows were observed to be restricted to seagrass beds; those found on unvegetated substrata were occupied by different alpheid shrimp – goby associations

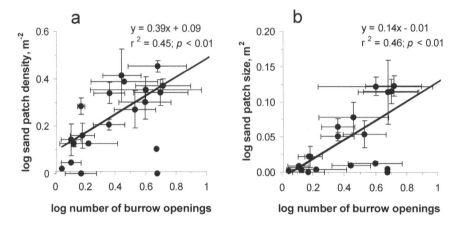

Figure 2. Significant linear relationships between disturbance marks mapped on the seagrass meadows. Error bars are standard errors of the mean (SEM).

explained only little of the total variance. Densities of sand patches were negatively correlated with shoot densities of *H. uninervis*; depth emerged as a negative correlate of shoot densities of *E. acoroides*.

Bolinao-Anda region (I): comparisons within seagrass sites and across habitats
Within the 20-km latitude covered by the survey, shrimp disturbance appeared to be more frequent in the northern beds of Santiago Island (Bolinao) than either midway, where turbid conditions existed, or in the beds of the reef flat of Cabarruyan Island (Anda) in the south (Fig. 1, a). When sites were regrouped *a posteriori* based on water clarity, the disturbance by alpheid shrimps became more pronounced in clear-water than in turbid seagrass areas (Fig. 3, top graphs). Sand mounds were similar in both environments, on average, but mound maxima, shaft densities, and shaft maxima were reduced in turbid areas (Fig 3, bottom graphs). The distribution patterns of shrimp disturbance were not similar, with sand patches occurring more regularly in the quadrats than the sand mounds and shafts, which occurred at random (Table 3). The distribution of burrow openings, on the other hand, appeared clumped (Table 3).

There was a clear overlap in the densities and sizes of sand patches and burrow openings of alpheid shrimps with those of sand mounds and of associated shafts of thalassinidean shrimps, especially on the shallow seagrass areas (~1 m depth) (Fig. 4, a). More sand mounds were also recorded on the bare substrates than in seagrass areas ($p < 0.05$; Fig. 4, b; Table 3), the latter having projected areas of 12 and 14% of our grid (1 m²) (maximum = 40 and 45%; maximum heights = 31 and 57 cm). The sizes of alpheid shrimp burrow openings were also significantly higher in the bare sediments than on seagrass beds ($p < 0.05$, Table 3; Fig. 4, b). We noted, however, that the unvegetated substrates had other shrimp-goby associations that accounted for the observed increase and not by the *A. macellarius–Cryptocentrus* sp. symbiosis that occurred in the seagrass beds. In Site 6 (Malilnep), for instance, we found occurrences of two banded *Alpheus* sp. individuals that alternated in expelling chela-loads of burrow sediment, and recognized the prawn goby *C. polyophthalmus* guarding several burrows.

Table 4. Variables that showed significant variation among regions (n=3) based on mixed analyses of variance versus *in-situ* bed characteristics (as covariates). All covariates were entered in each analysis but only the significant ones* are presented below (error term in this case is 'within sites'; α =0.05). The association of the covariables with the dependent variable is given as + or −.

Dependent variable	p Region	% variance explained by Region	Covariables	Error	Covariables in model		% variance explained	p
Sand patch density (1)	0.043	1	91	9	Burrow opening density	+	3	<0.001
					Sand mound density	−	1	0.003
					Shaft/ pit density	+	<1	0.048
					H. uninervis shoot density	−	<1	0.042
Sand mound size (2)	<0.001	9	87	13	Depth	+	2	0.003
					Sand patch size	−	1	0.019
					Shaft/ pit opening size	+	2	0.002
					Sand mound density	+	4	<0.001
					T. hemprichii shoot density	−	4	<0.001
n seagrass species (3)	<0.001	2	96	96	Skewness	+	1	0.018
Thalassia hemprichii shoot density (4)	<0.001	20	85	15	Sand patch density	+	1	0.020
					Sand patch size	−	1	0.026
					Sand mound density	+	2	0.016
					Sand mound size	−	4	<0.001
					Shaft/ pit opening sie	+	1	0.026
Halodule uninervis shoot density (5)	0.024	7	60	40	Sand patch density	−	3	0.047
					H. ovalis shoot density	+	11	<0.001
Enhalus acoroides shoot density (6)	0.027	3	83	17	Depth	−	2	0.007

Note: Separate stepwise regressions included the following covariables as predictors of the dependent variables, but altogether explained much less variance than that accounted for by ANOVA: (1) sediment skewness (+) in place of *H. uninervis* shoot density; (2) sand patch density (−) and *C. rotundata* shoot density (+) in place of *T. hemprichii* shoot density; (3) sand patch density (+) and mud content (−); (4) mean grain size (−); (5) burrow opening density in place of sand patch density; and (6) mud content (−) and sediment sorting (−).

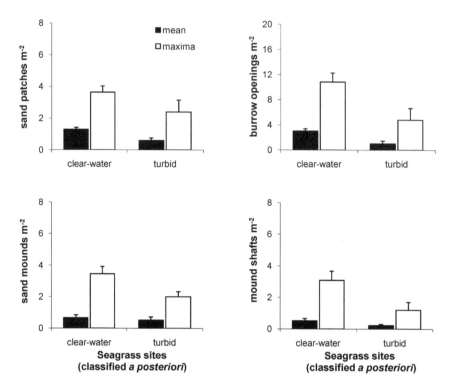

Figure 3. Region 1, Bolinao-Anda: Differences in the frequencies of shrimp disturbance between clear-water (n=11) and turbid (n=5) seagrass areas. Bars are means (■) or maxima (□) ± SEM.

Vertical profiles of sediment properties

The profiles of undisturbed seagrass sediments were clearly different between Sites 5 (Silaki) and 12 (Rufina). Mean grain sizes were between 1.0 and 1.5ϕ in Site 5 while the bed sediment in Site 12 was more heterogeneous, i.e., μ between 0.1 and 1.25ϕ, and with coarser grains from 12 cm and deeper. Both burrowing alpheid and thalassinidean shrimps altered sediment texture but the effect of the shrimps was significant only in Site 5 (2-way ANOVA, $p < 0.005$), where the redistribution of coarser grain sizes was evident (Fig. 5). The vertical heterogeneity of the bed substrate in Site 12 appeared to obscure the effect of the shrimps in redistributing coarse grains in the upper 10 cm and in consolidating fine fractions in deeper sediments (2-way ANOVA, $p > 0.05$; Fig. 5).

Organic matter (OM) content in the undisturbed seagrass sediments was statistically similar for both sites and averaged ~40 ± 2 mg•g DW^{-1}. Alpheid shrimps increased OM only slightly (41 ± 1.3 mg•g DW^{-1}) while thalassinidean shrimps significantly raised OM content by 20% (50 ± 1 mg•g DW^{-1}, Fig. 6, a; 2-way ANOVA, $p < 0.001$). Total N profiles (as NH_4-N) had significant differences between the sites and concentrations were reduced by up to two-fold at the top 10 cm of shrimp-disturbed sediments (Fig. 6, b). In Site 5, shrimp disturbance seemed to keep nitrogen levels lower (0.24 ± 0.01 mg NH4-N•g DW^{-1}) than in the undisturbed sediments (0.33 ± 0.01 mg NH_4-N•g DW^{-1}) (Tukey's HSD, $p < 0.017$). Nitrogen was significantly higher in the top 10 cm layer than deeper in the

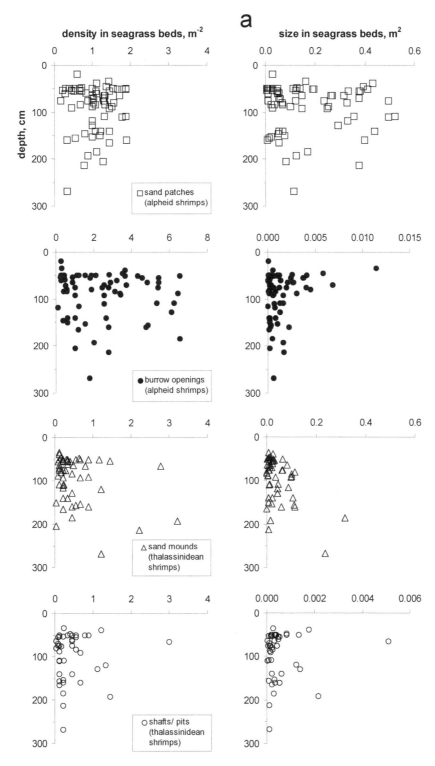

Figure 4. Bolinao-Anda region (I) – Distributions of burrowing shrimp disturbance marks with sampling depth in (a) seagrass meadows and (b) unvegetated substrata.

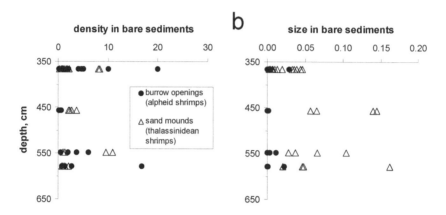

Figure 4 (continued)

undisturbed seagrass sediments of Site 12 (Tukey's HSD, $p < 0.017$; Fig. 6, b); shrimp disturbance only reduced nitrogen levels on the top 10 cm of the sediments by 20-73% and by 4-36% at deeper than 10 cm, but did not affect overall profiles. Nitrogen levels of disturbed sediments in Site 12 also exhibited a positive linear relationship with OM ($y = 0.003x - 0.002$, $r^2 = 0.35$, $p < 0.001$), with low levels associated with alpheid shrimps and the high ones with the thalassinidean shrimps. The sites also differed in total phosphorus concentrations, which were higher in Site 12 (3.75 ± 0.12 mg·g o-PO$_4$·g DW^{-1}) (Fig. 6, c) than in Site 5 (3.02 ± 0.01 mg o-PO$_4$·g DW^{-1}) (3-way ANOVA, $p < 0.05$), and shrimp disturbance did not influence either concentration or distribution of phosphorus in the sediments (3-way ANOVA, $p > 0.05$) (Fig. 6, c).

Discussion

The primary agents of readily observable disturbance in the seagrass beds studied were alpheid shrimps, *Alpheus macellarius*, which associated with at least two species of prawn-goby *Cryptocentrus* (Palomar et al. 2004), i.e., the blue-speckled prawn goby *C. octafasciatus* Regan, 1908 and the Singapore prawn goby *C. singapurensis* (Herre, 1936). The observed restricted distribution imposed by vegetation suggests the animal's food habit, thus, leaf clipping is an apparent behaviour for *A. macellarius* on the meadows (Nacorda et al., Chapter 3, this Thesis). Outside the seagrass beds, other distinct *Alpheus–Cryptocentrus* combinations were noted. Examples of such pairs frequently reported in sandy substrates within the region include *A. bellulus* Miya and Miyake, 1969 – *Cryptocentrus cinctus* (Herre, 1936), *A. bellulus–Cryptocentrus* sp. 1, *A. djiboutensis* De Man, 1909 – *C. caeruleomaculatus* (Rupell, 1830), and *A. djiboutensis–C. singapurensis* (Nakasone and Manthachitra 1986, Manthachitra and Sudara 1988). Densities of burrow openings were less than those reported by Bradshaw (1997) from reefal sediments of Phuket.

The next important disturbance agents were pooled as a distinct taxon group, the thalassinidean shrimps, in view of the difficulty in obtaining specimens for identification, and because we assumed that the relative effects on the meadows are

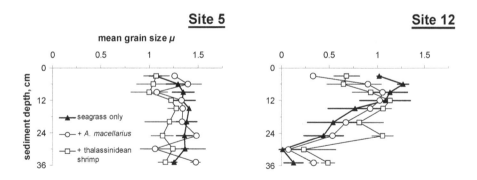

Figure 5. Vertical distribution profiles of mean grain size μ (as ϕ, ± SEM) in undisturbed (▲) and shrimp-disturbed sediments of Sites 5 (Silaki) and 12 (Rufina) (Bolinao-Anda region).

independent of the exact species. Hence, their distribution was not restricted to the vegetated substrates and the shrimps were common at greater depths below the depth limit of seagrass beds and in non-vegetated lagoon bottoms of the Bolinao-Anda region. The sand mound densities we found on seagrass and bare substrates correspond to the range of 'low-' and 'medium-density areas' reviewed by de Vaugelas (1985) and are within the range observed in shallow backreef lagoons (Roberts et al. 1981) and reefal sediments (Bradshaw 1997). We did not find any significant correlation between the densities of these two dominant burrowing shrimps, suggesting that other factors than interspecific competition for space may operate to constrain their densities. In addition, the observed variation was considerable.

Hydrodynamics appeared to influence the occurrence of burrowing shrimps. The disturbance of alpheid and thalassinidean shrimps was a constant feature of developed meadows with benign to mid-range hydrodynamics. Beds exposed to higher water movements have less or with shifting coarse sediments, e.g., where *Thalassodendron ciliatum* occurs (Bandeira 2002) or where a dense mix of species is closely integrated on hard bottoms. Seagrasses were observed to adapt to such environments by allocating considerable resources to vertical growth (Duarte et al. 1996), growing longer rhizome internodal distances (Nathaniel et al. 2003) and maintaining higher belowground biomass, which could be unsuitable for shrimp bioturbation (Townsend and Fonseca 1998). However, when disturbance exposes large gaps within these beds, e.g., the 1,200 m²-gap of thin sediments in Bolinao in 1999 (Olesen et al. 2004), we observed burrowing alpheid shrimps to immediately emerge and transport sediments to the surface (pers. obs). The animals seemed opportunistic – these were smaller than usually found in other beds and their burrow openings were densely packed (up to 40 m⁻²), perhaps until the large gap completely became revegetated. In other relatively wave-exposed areas of meadows (e.g., Sites 3 and 18), bioturbation pits of the thalassinidean shrimp *Corallianassa* sp. were evident. These ghost shrimps do not actively rework the sediment and use their burrow openings to catch drifting seagrass/ detrital fragments and store them as cache (Griffis and Suchanek 1991, Dworschak et al. 2006).

Water clarity emerged as a significant factor for densities of *A. macellarius* disturbance (sand patches, burrow openings). The abundance of *A. macellarius* in

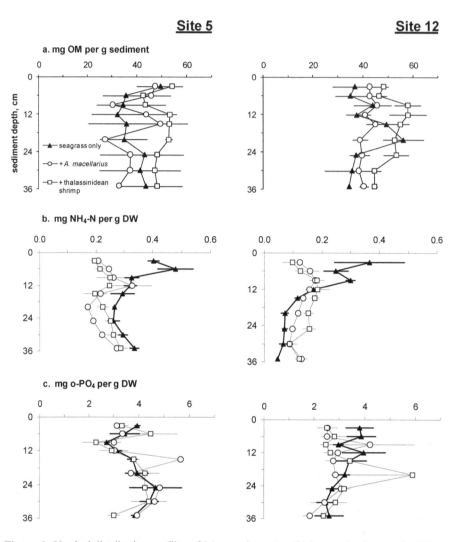

Figure 6. Vertical distribution profiles of (a) organic matter, (b) inorganic nitrogen (as NH₄-N), and (c) inorganic phosphorus (as o-PO₄) contents in the undisturbed and shrimp-disturbed sediments of Sites 5 and 12 in the Bolinao-Anda region. Data are means ± SEM.

clear-water meadows is linked to their prawn gobies' camouflage and ability to perceive risk factors (e.g., predation) in the lighted environment (Karplus 1987). Because this view of the ambient environment becomes limited in turbid waters, the gobies could be trading-off hiding with the perceived risky emergence at the sediment surface. Also, since sediment type in silty beds are less compact and resuspend easily, shrimp burrows could be expectedly unstable and the openings would easily be filled up with sediments. These may cause the animal partners to stay hidden from view and the shrimp spending more burrow reconstruction and maintenance time rather than extending the burrows or harvesting seagrass leaves, at least until conditions become more favorable for emergence at the surface.

The significant correlations of shrimp gap attributes with sediment and seagrass parameters explained only 1 to 3% of variability in the ANOVAs (Table 3). However, the significant relationship between sand patch density and number of seagrass species, despite large variation in the data (Table 4), offers a mechanism by which small colonizing species are maintained in mixed meadows (Duarte et al. 1997). As much as 14% of the meadow on average represents an open space for *Halophila ovalis*, noticeable on the sand mounds and capable of fast clonal expansion (Vermaat et al. 1995, Rasheed 2004). Once recolonized, the unconsolidated sediments become stabilized (Fonseca 1989), hence, prepared for subsequent colonizers, and the new vegetation becomes functional, e.g., as food, refuge from predators, and habitat (Bell and Westoby 1986). This series of processes for fast colonizers was thought to minimize ecosystem function losses due to frequent small-scale disturbance (Duarte 2000). Sediment burial reduces the cover and productivity of seagrasses (e.g., at >10 mounds m^{-2}, Roberts et al. 1981; but see also Chapter 5, this Thesis) but was also shown to elicit species-specific responses to cope with such disturbance, e.g., increased vertical growth, branching, and shoot densities (Duarte et al. 1997). *Cymodocea nodosa* grew upwards from rhizomes buried after the passage of sand waves (Marbá et al. 1994) while *Zostera novazelandica* moderated the effect of high sediment turnover activity by *Callianassa filholi* by higher above-ground plant growth during the warmer months (Berkenbusch et al. 2000).

Bioturbation by burrowing shrimps, as an example of case 4 allogenic engineering (Jones et al. 1994), has major consequences on local bed sedimentology and geochemistry, interface processes, the fate of benthic communities, and, eventually, on the rhizosphere of seagrass beds. Larger grain sizes were more prominent in the upper 10 cm section of sediments than deeper in the shrimp-disturbed cores (Fig. 5). Both types of shrimps rearrange large grains differently: thalassinidean shrimps bury such fractions whereas alpheid shrimps bring them to the surface (Bradshaw and Scoffin 2001). Rearrangements in the sediment framework results in redistribution and continual transfer of sediment organic matter and nutrients, which affects biogeochemical redox processes (Schulz 2000). The reworked but unconsolidated sediments are subject to resuspension (Aller and Dodge 1974, Rowden et al. 1998) and the associated benthos may either be enhanced (e.g., for bacteria; Gilbert et al. 1998) or diminished due to smothering and/ or burial, e.g., for diatoms, meiofauna (Suchanek and Colin 1986, Branch and Pringle, 1987), and macrofauna (Posey et al. 1991, Berkenbusch et al. 2000). The presence of complex burrows that open to the sediment-water interface provides oxygenation of the sediments (Forster and Graf 1992) and bioventilation redistributes oxygen and nutrients to the water column (Koike and Mukai 1983, Waslenchuk et al. 1983, Murphy and Kremer 1992, Forster 1996). Oxygenation alleviates the moderately reducing conditions of the sediments (Terrados et al. 1999) and disturbance could be a mechanism to diffuse oxygen from seagrass roots (Hemminga 1998, Pedersen et al. 1998, Frouin 2000), hence, help accelerate organic matter degradation (Ziebis et al. 1996).

Acknowledgments. This study was funded by WOTRO (WB84-413) and partially supported by the DOST/PCAMRD *Marine Science Program at the Kalayaan Islands* Project and the UP-CIDS Project *Comprehensive and Sustainable*

Development Program of Lopez Jaena, Misamis Occidental. We thank Jack Rengel, Criz Ragos, and Ronald de Guzman for help in the field and laboratory, and the generous assistance of our numerous volunteers – Andrea and Jonathan Persons, Dr. Sai Thampanya, Anjo Tiquio, Alvin Carlos, Day Lacap, Dr. Rev Molina, Jhun Castrence, Dr. Wili Uy, Freddie Lagarde, Lt. J. Bacordo, Karen Araño, and Fe Pillos-Lomahan. Dr. Charles Fransen identified the species of *Alpheus*. Dr. Rene Rollon and Ines Templo shared data from the cruise to the Tubbataha Reefs. Dr. Fred Kruis accommodated HEN in the lab at IHE and granted technical and practical advice on the methods; Ingrid de Bont provided unwavering help in the sediment nutrient analyses. Dr. Al Licuanan gave suggestions on interpreting spatial patterns and local statistics.

References

Aller, R.C. and Dodge, R.E. 1974. Animal-sediment relations in a tropical lagoon, Discovery Bay, Jamaica. Journal of Marine Research 32: 209-232.

Arriesgado, D.M. 1999. Population dynamics of *Thalassia hemprichii* (Ehrenb.) Aschers. in the coastal waters of Punta Sulawan, Tubajon, Laguindingan, Misamis Oriental. M.Sc. Thesis, Mindanao State University-Naawan, School of Marine Fisheries and Technology, Misamis Oriental, Philippines.

Bach, S.S., Borum, J., Fortes M.D., and Duarte, C.M. 1998. Species composition and plant performance of mixed seagrass beds along a siltation gradient at Cape Bolinao, The Philippines. Marine Ecology Progress Series 174: 247-256.

Bandeira, S.O. 2002. Leaf production rates of *Thalassodendron ciliatum* from rocky and sandy habitats. Aquatic Botany 72: 13-24.

Barnes, D.J. and Lough, J.M. 1999. *Porites* growth characteristics in a changed environment: Misima Island, Papua New Guinea. Coral Reefs 18: 213-218.

Bell J.D. and Westoby M. 1986. Abundance of macrofauna in dense seagrass is due to habitat preference, not predation. Oecologia 68: 205-209.

Berkenbusch, J., Rowden, A.A., and Probert, P.K. 2000. Temporal and spatial variation in macrofauna community composition imposed by ghost shrimp *Callianassa filholi* bioturbation. Marine Ecology Progress Series 192: 249-257.

Bradshaw, C. 1997. Bioturbation of reefal sediments by crustaceans in Phuket, Thailand. Proceedings of the 8th International Coral Reef Symposium 2: 1801-1806.

Bradshaw, C. and Scoffin, T.P. 2001. Differential preservation of gravel-sized bioclasts in alpheid- versus callianassid-bioturbated muddy reefal sediments. Palaois 16: 185-191, DOI: 10.1669/0833-1351(2001)016<0185:DPOGSB>2.0.CO;2.

Branch, G.M. and Pringle, A. 1987. The impact of the sand prawn *Callianassa kraussi* Stebbing on sediment turnover, and on bacteria, meiofauna and benthic microflora. Journal of Experimental Marine Biology and Ecology 107: 219-235.

Buchanan, J.B. 1984. Sediment analysis. In: Holme, N.A. and McIntyre, A.D., eds., Methods for the study of marine benthos, pp. 41-65. Blackwell Scientific Publications, Philadephia.

Chapin, F.S. III, Walker, B.H., Hobbs, R.J., Hooper, D.U., Lawton J.H., Sala, O.E., and Tilman, D. 1997. Biotic control over the functioning of ecosystems. Science 277: 500-503.

Dawes, C.J., Andorfer, J., Rose, C., Uranowski, C., and Ehringer, N. 1997. Regrowth of the seagrass *Thalassia testudinum* into propeller scars. Aquatic Botany 59: 139-155.

De Iong, H.H., Wenno, B., and Meelis, E. 1995. Seagrass distribution biomass changes in relation to dugong grazing in the Moluccas, Indonesia. Aquatic Botany 50: 1-19.

Duarte, C.M. 2000. Marine biodiversity and ecosystem services: an elusive link. Journal of Experimental Marine Biology and Ecology 250: 117-131.

Duarte, C.M., Hemminga, M.A., and Marbá, N. 1996. Growth and population dynamics of *Thalassodendron ciliatum* in a Kenyan back-reef lagoon. Aquatic Botany 55: 1-11.

Duarte, C.M., Terrados, J., Agawin, N.S.R., Fortes, M.D., Bach, S., and Kenworthy, W.J. 1997. Response of a mixed Philippine seagrass meadow to experimental burial. Marine Ecology Progress Series 147: pp. 285-294.

Dumbauld B.R. and Wyllie-Echeverria, S. 2003. The influence of burrowing thalassinid shrimps on the distribution of intertidal seagrasses in Willapa Bay, Washington, USA. Aquatic Botany 77: 27-42.

Dworschak, P. and Ott, J.A. 1993. Decapod burrows in mangrove-channel and back-reef environments at the Atlantic Barrier Reef, Belize. Ichnos 2: 277-290.

English, S., Wilkinson, C. and Baker, V., eds. 1994. Survey manual for tropical marine resources. Australian Institute of Marine Science, Townsville, Australia, 368 p.

Erftemeijer, P.L.A., Drossaert, W.M.E., and Smekens M.J.E. 1993. Macrobenthos of two contrasting seagrass habitats in South Sulawesi, Indonesia. Wallaceana 70: 5-12.

Fonseca M.S. 1989. Sediment stabilization by *Halophila decipiens* in comparison to other seagrasses. Estuarine, Coastal and Shelf Science 29: 501-507.

Fonseca, M.S. and Bell, S.S. 1998. Influence of physical setting on seagrass landscapes near Baeufort, North Carolina, USA. Marine Ecology Progress Series 17: 109-121.

Fonseca, M.S., Zieman, J.C., Thayer, G.W., and Fisher, J.S. 1983. The role of current velocity in structuring eelgrass (*Zostera marina* L.) meadows. Estuarine, Coastal and Shelf Science 17: 367-380.

Forster, S. 1996. Spatial and temporal distribution of oxidation events occurring below the sediment-water interface. P.S.Z.N. I: Marine ecology 17: 309-319.

Forster, S. and Graf, G. 1992. Continuously measured changes in redox potential influenced by oxygen penetrating from burrows of *Callianassa subterranea*. Hydrobiologia 235/236: 527-532.

Francour, P., Ganteaume, A., and Poulain, M. 1999. Effects of boat anchoring in *Posidonia oceanica* seagrass beds in the Port-Cos National Park (northwestern Mediterranean Sea). Aquatic Conservation: Marine and Freshwater Ecosystems 9: 391-400.

Frouin, P. 2000. Effects of anthropogenic disturbances of tropical soft-bottom benthic communities. Marine Ecology Progress Series 194: 39-53.

Gilbert, F., Stora, G., and Bonin, P. 1998. Influence of bioturbation on denitrification activity in Mediterranean coastal sediments: an *in situ* experimental approach. Marine Ecology Progress Series 163: 99-107.

Griffis, R.B. and Suchanek, T.H. 1991. A model of burrow architecture and trophic modes in thalassinidean shrimps (Decapoda: Thalassinidea). Marine Ecology Progress Series 79: 171-183.

Hall, S.J., Raffaelli, D., and Thrush, S.F. 1992. Patchiness and disturbance in shallow water benthic assemblages. In: Giller, P.S., Hildrew, and A.G. Raffaelli, D.G., eds., Aquatic ecology: scale, pattern and process, Proceedings of the British Ecological Society and the American Society of Limnology and Oceanography Symposium, pp. 333-375. Blackwell Scientific Publications, Oxford.

Hastings, K., Hesp, P., and Kendrick, G.A. 1995. Seagrass loss associated with boat moorings at Rottnest Island, Western Australia. Ocean and Coastal Management 26: 225-246.

Heck, K.L. and Valentine, J.F. 1995. Sea urchin herbivory: evidence for long-lasting effects in subtropical seagrass meadows. Journal of Experimental Marine Biology and Ecology 189: 205-217.

Hemminga, M.A. 1998. The root/ rhizome system of seagrasses: an asset and a burden. Journal of Sea Research 39: 183-196.

Hemminga, M.A. and Duarte, C.M. 2000. Seagrass ecology. Cambridge University Press, 298 p.

Jacobs, R.P.W.M., den Hartog, C., Braster, B.F., and Carrier, F.C. 1981. Grazing of the seagrass *Zostera noltii* by birds at Terschelling (Dutch Wadden Sea). Aquatic Botany 10: 241-259.

Jennings, S. and Kaiser, M.J. 1998. The effects of fishing on marine ecosystems. Advances in Marine Biology 34: 201-352.

Jones, C.G., Lawton, J.H., and Shachak, M. 1994. Organisms as ecosystem engineers. Oikos 69: 373-386.

Kamp-Nielsen, L., Vermaat, J.E., Wesseling, I., Borum, J., and Geertz-Hansen, O. 2002. Sediment properties along gradients of siltation in South-east Asia. Estuarine, Coastal and Shelf Science 54: 127-137.

Karplus, I. 1987. The association between gobiid fishes and burrowing alpheid shrimps. Oceanography and Marine Biology Annual Reviews 25: 507-562.

Kenworthy, W.J., Fonseca, M.S., Whitfield, P.E., and Hammerstrom, K. 2000. Experimental manipulation and analysis of recovery dynamics in physically disturbed tropical seagrass communities of North America: implications for restoration and management. Biologia Marina Mediterranea 7: 385-388.

Koike, I. and Mukai, H. 1983. Oxygen and inorganic contents and fluxes in burrows of the shrimps *Callianassa japonica* and *Upogebia major*. Marine Ecology Progress Series 12: pp. 185-190.

Kruis, F. 2000. Environmental chemistry: selected analytical methods. IHE-Delft, The Netherlands, 84 p.

Leeder, M.R. 1982. Grain properties. In: Sedimentology, pp. 35-43. George Allen and Unwin, London.

Manthachitra, V. and Sudara, S. 1988. Behavioral communication patterns between gobiid fishes and alpheid shrimps. In: Choat, J.H., Barnes, D., Borowitzka, M.A., Coll, J.C., Davies, P.J., Flood, P., Hatcher, B.G., Hopley, D., Hutchings, P.A., Kinsey, D., Orme, G.R., Pichon, M., Sale, P.F., Sammarco, P., Wallace, C.C., Wilkinson, C., Wolanski, E., and Bellwood, O., eds., Proceedings of the 6th International Coral Reef Symposium: Vol. 2: Contributed Papers, pp. 769-773. Australian Institute of Marine Science, Townsville, Australia.

Marbá, N., Cebrián, J., Enríques S., and Duarte, C.M. 1994. Migration of large-scale subaqueous bed forms measured with seagrasses (*Cymodocea nodosa*) as tracers. Limnology and Oceanography 39: 126-133.

McClanahan, T.R., Obura, D., 1997, Sedimentation effects on shallow coral communities in Kenya. Journal of Experimental Marine Biology and Ecology 209: 103-122

McManus, J.W., Nañola, C.L. Jr., Reyes, R.B. Jr., and Kesner, K.N. 1992. Resource ecology of the Bolinao coral reef system. ICLARM Studies and Reviews 22, 117 p. International Center for Living Aquatic Resources Management, Manila, Philippines.

McManus, L.T. and Chua, T.-E. 1990. The coastal environmental profile of Lingayen Gulf, Philippines. ICLARM Technical Report 22, 69 p. International Center for Living Aquatic Resources Management, Manila, Philippines.

Milliman, J. and Meade, R.J. 1983. World-wide delivery of river sediments to the oceans. Journal of Geology 91: 1-21.

Murphy, R.C. and Kremer, J.N. 1992. Benthic community metabolism and the role of deposit-feeding callianassid shrimp. Journal of Marine Research 50: 321-340.

Nacken, M. and Reise, K. 2000. Effects of herbivorous birds on intertidal seagrass beds in the northern Wadden Sea. Helgoland Marine Research 54: 87-94.

Nacorda, H.M.E., Stamhuis, E.J., and Vermaat, J.E. Chapter 3: Aboveground behavior and significance of *Alpheus macellarius*, Chace, 1988, in a Philippine seagrass meadow. (this Thesis)

Nacorda, H.M.E., Stamhuis, E.J., and Vermaat, J.E. Chapter 4: Burrows and behaviour of the snapping shrimp *Alpheus macellarius*, Chace, 1988, in different seagrass substrates. (this Thesis)

Nakasone Y. and Manthachitra V. 1986. A preliminary report on the associations between gobies and alpheid shrimps in the Sichang Island, the Gulf of Thailand. Galaxea 5: 157-162.

Nickell, L.A. and Atkinson, R.J.A. 1995. Functional morphology of burrows and trophic modes of three thalassinidean shrimp species, and a new approach to classification of the thalassinidean burrow morphology. Marine Ecology Progress Series 128: 181-197.

Ogden, J.C., Brown, R.A., and Salesky, N. 1973. Grazing by the echinoid *Diadema antillarum* Philippi: formation of halos around West Indian patch reefs. Science 182: 715-717.

Olesen, B., Marbá, N., Duarte, C.M., Savela, R.S., and Fortes, M.D. 2004. Recolonization dynamics in a mixed seagrass meadow: the role of clonal versus sexual processes. Estuaries 27: 770-780.

Palomar, N.E., Juinio-Meñez, M.A., and Karplus, I. 2004. Feeding habits of the burrowing shrimp *Alpheus macellarius*. Journal of the Marine Biological Association of the United Kingdom 84: 1199-1202.

Palomar, N.E., Juinio-Meñez, M.A., and Karplus, I. 2005. Behavior of the burrowing shrimp *Alpheus macellarius* in varying gravel substrate conditions. Journal of Ethology 23: 173-180.

Patriquin, D.G. 1975. "Migration" of blowouts in seagrass beds at Barbados and Carriacou, West Indies, and its ecological and geological implications. Aquatic Botany 1: 163-189.

Pedersen, O., Borum, J., Duarte, C.M., and Fortes, M.D. 1998. Oxygen dynamics in the rhizosphere of *Cymodocea rotundata*. Marine Ecology Progress Series 169: 283-288.

Philippart, C.J.M. 1994. Interactions between *Arenicola marina* and *Zostera noltii* on a tidal flat in the Dutch Wadden Sea. Marine Ecology Progress Series 111: 251-257.

Posey, M.H., Dumbauld, B.R., and Armstrong, D.A. 1991. Effects of a burrowing mud shrimp, *Upogebia pugettensis* (Dana), on abundances of macro-infauna. Journal of Experimental Marine Biology and Ecology 148: 283-294.

Preen, A. 1995. Impacts of dugong foraging on seagrass habitats: observational and experimental evidence for cultivation grazing. Marine Ecology Progress Series 24: 201-213.

Preen, A.R., Lee Long, W.J., and Coles, R.G. 1995. Flood and cyclone related loss, and partial recovery of more than 1000 km^2 of seagrass in Hervey Bay, Queensland, Australia. Aquatic Botany 52: 3-17.

Rasheed, M.A. 2004. Recovery and succession in a multi-species tropical seagrass meadow following experimental disturbance: the role of sexual and asexual reproduction. Journal of Experimental Marine Biology and Ecology 30: 13-45.

Rivera, P.C. 1997. Hydrodynamics, sediment transport and light extinction off Cape Bolinao, Philippines. PhD Thesis, Wageningen Agicultural University and IHE-Delft. A.A. Balkema, Rotterdam, 244 p.

Roberts, H.H., Wiseman, W.J. Jr., and Suchanek, T.H. 1981. Lagoon sediment transport: the significant effect of *Callianassa* bioturbation. Proceedings of the 4th International Coral Reef Symposium, Manila 1: 459-465. Marine Science Institute, University of the Philippines, Diliman, Quezon City, Philippines.

Rogerson, P.A. 2001. Statistical methods for geography. Sage Publications, London, 236 p.

Rollon, R.N. 1998. Spatial variation and seasonality in growth and reproduction of *Enhalus acoroides* (L.f.) Royle populations in the coastal waters off Cape Bolinao, NW Philippines. PhD Thesis, Wageningen Agricultural University and IHE-Delft. A.A. Balkema, Rotterdam, The Netherlands, 135 p.

Rose, C.D., Sharp W.C., Kenworthy W.J., Hunt J.H., Lyons W.G., Prager, E.J., Valentine, J.F., Hall, M.O., Whitfield P.E., and Fourqrean, J.W. 1999. Overgrazing of a large seagrass bed by the sea urchin *Lytechinus variegatus* in Outer Florida Bay. Marine Ecology Progress Series 190: 211-222.

Rowden, A.A., Jones, M.B., and Morris, A.W. 1998. Sediment turnover estimates for the mud shrimp *Callianassa subterranea* (Montagu) (Thalassinidea) and its influence upon resuspension in the North Sea. Continental Shelf Research 18: 1365-1380.

Schulz, H.D. 2000. Quantification of early diagenisis: dissolved constituents in marine pore water. In: Schulz H.D. and Zabel, M, eds., Marine geochemistry, pp.85-128. Springer, Berlin.

Sheppard, J.K., Lawler, I.R., and Marsh, H. 2007. Seagrass as pasture for seacows: landscape-level dugong habitat evaluation. Estuarine, Coastal and Shelf Science 71: 117-132.

Sokal, R.R. and Rohlf, F.J. 1995. Biometry, the principles and practice of statistics in biological research, 3rd ed. WH Freeman and Co, New York, 887 p.

Stafford, N. and Bell, S. 2003. Biological correlates of hydrodynamic regime in a seagrass landscape. Poster presented at the 88th Annual Meeting of The Ecological Society of America held jointly with the International Society for Ecological Modeling – North American Chapter, August 3-8, 2003, Savannah, Georgia (http://abstracts.co.allenpress.com/pweb/esa2003/document/?ID=27050)

Stamhuis, E.J., Reede-Dekker, T., van Etten, Y., de Wiljes, J.J., and Videler, J.J. 1996. Behaviour and time allocation of the burrowing shrimp *Callianassa subterranea*

(Decapoda, Thalassinidea). Journal of Experimental Marine Biology and Ecology 204: 225-239.

Stapel, J. and Erftemeijer, P.L.A. 1997. Leaf harvesting and sediment reworking by burrowing alpheid shrimps in a *Thalassia hemprichii* meadow in South Sulawesi, Indonesia. In: Stapel, J., Nutrient dynamics in Indonesian seagrass beds: factors determining conservation and loss of nitrogen and phosphorus, pp. 33-41. PhD Thesis, Katholieke Universiteit Nijmegen. WOTRO/ NWO, 127 p.

Suchanek, T.H. 1983. Control of seagrass communities and sediment distribution by *Callianassa* (Crustacea, Thalassinidea) bioturbation. Journal of Marine Research 41: 281-298.

Suchanek, T.H. and Colin, P.L. 1986. Rates and effects of bioturbation by invertebrates and fishes at Enewetak and Bikini Atolls. Bulletin of Marine Science 38: 25-34.

Terrados, J., Duarte, C.M., Fortes, M.D., Borum, J., Agawin, N.S.R., Bach, S., Thampanya, U., Kamp-Nielsen, L., Kenworthy, W.J., Geertz-Hansen, O., and Vermaat, J. 1998. Changes in community structure and biomass along gradients of siltation in SE Asia. Estuarine, Coastal and Shelf Science 46: 757-768.

Terrados, J., Duarte, C.M., Kamp-Nielsen, L., Borum, J., Agawin, N.S.R., Fortes, M.D., Gacia, E., Lacap, D., Lubanski, M., and Greve, T. 1999. Are seagrass growth and survival affected by reducing conditions in the sediment? Aquatic Botany 65: 175-197.

Townsend E.C. and Fonseca, M.S. 1998. Bioturbation as a potential mechanism influencing spatial heterogeneity of North Carolina seagrass beds. Marine Ecology Progress Series 169: 123-132.

Uy, W.H., Hemminga, M.A., Vermaat, J.E., and Fortes, M.D. 2001. Growth, morphology and photosynthetic responses of *Thalassia hemprichii* and *Halodule uninervis* to long-term *in situ* light reduction. In: Uy, W.H., Functioning of Southeast Asian seagrass species under deteriorating light conditions, pp. 9-33. PhD Thesis, Wageningen University and IHE-Delft. Swets and Zeitlinger, Lisse, 121 p.

Valentine, J.F., Heck, K.L. Jr., Harper, P., and Beck, M. 1994. Effects of bioturbation in controlling turtlegrass (*Thalassia testudinum* Banks *ex* König) abundance: evidence from field enclosures and observations in the Northern Gulf of Mexico. Journal of Experimental Marine Biology and Ecology 178: 181-192.

Van Katwijk, M.M., Meier, N.F., van Loon, R., von Hove, E.M., Giesen, W.B.J.T., van der Velde, G., and den Hartog, C. 1993. Sabaki River sediment load and coral stress: correlation between sediments and condition of the Malindi-Watamu reefs in Kenya (Indian Ocean). Marine Biology 117: 675-683

Vaugelas, J.D. 1985. Sediment reworking by callianassid mud-shrimp in tropical lagoons: a review with perspectives. Proceedings of the 5th International Coral Reef Congress, Tahiti, 6: 617-622.

Vaugelas, J.D. and Buscail, R. 1990. Organic matter distribution in burrows of the thalassinid crustacean *Callichirus laurea*, Gulf of Aqaba (Red Sea). Hydrobiologia 207: 269-277.

Vermaat, J.E., Agawin, N.S.R., Duarte, C.M., Fortes, M.D., Marbá, N., and Uri, J.S. 1995. Meadow maintenance, growth and productivity of a mixed Philippine seagrass bed. Marine Ecology Progress Series 124: 215-225.

Walker, D.I., Lukatelich, R.J., Bastyan, G., and McComb, A.J. 1989. Effect of boat moorings on seagrass beds near Perth, Western Australia. Aquatic Botany 36: 69-77.

Waslenchuk, D.G., Matson, E.A., Zajac, R.N., Dobbs, F.C., and Tramontano, J.M. 1983. Geochemistry of burrow waters vented by a bioturbating shrimp in Bermudan sediments. Marine Biology 72: 219-225.

Wesseling, I., Uychiaoco, A.J., Aliño, P.M., and Vermaat, J.E. 2001. Partial mortality in *Porites* corals: variation among Philippine reefs. International Review of Hydrobiology 86: 77-85

Woods, C.M.C. and Schiel, D.R. 1997. Use of seagrass *Zostera novazelandica* (Setchell, 1933) as habitat and food by the crab *Macrophthalmus hirtipes* (Heller, 1862) (Brachyura, Ocypodidae) on rocky intertidal platforms in southern New Zealand. Journal of Experimental Marine Biology and Ecology 214: 49-65.

Ziebis, W., Forster, S,, Huettel, M. and Jørgensen, B.B. 1996. Complex burrows of the mud shrimp *Callianassa truncata* and their geological impact in the seabed. Nature 382: 619-622.

Chapter 3

Aboveground behavior of the snapping shrimp *Alpheus macellarius*, Chace, 1988, and its significance for leaf and nutrient turnover in a Philippine seagrass meadow

H.M.E. Nacorda[1, 2], E.J. Stamhuis[3], J.E. Vermaat[2, 4]

[1] Marine Science Institute, University of the Philippines, UPPO Box 1, Diliman, Quezon City, 1101 The Philippines

[2] Department of Environmental Resources, UNESCO-IHE Institute for Water Education, PO Box 3015,2601 DA Delft, The Netherlands

[3] Department of Marine Biology, University of Groningen, PO Box 14, 9750 AA Haren, The Netherlands

[4] present address: Institute for Environmental Studies, Vrije Universiteit, De Boelelaan 1087, 1081 HV Amsterdam, The Netherlands

Abstract

The aboveground behavior of the snapping shrimp *Alpheus macellarius*, Chace, 1988, was studied in a shallow clear–water seagrass meadow off Bolinao, NW Philippines. The shrimp's activity pattern was analyzed from short video records taken every hour on sampling visits during the dry and wet months. *Alpheus macellarius* actively moved sediment (frequency = 77 h^{-1}), often surveyed its burrow opening (21 h^{-1}), stacked up rubble (17 h^{-1}), and occasionally harvested seagrass leaves (9 h^{-1}) during the dry months. Time allocated for these aboveground activities was only 12% of its daily active period of 9 hours during daylight; *A. macellarius* was in its burrow during the remainder of daytime. During the wet months, the shrimp's activity rates and corresponding time allocations were reduced by at least 34% that of the dry months and its within-burrow period consequently increased by 5%. Overall, sediment moving and harvesting activities of *A. macellarius* contributed frequent disturbance to the meadow: an individual shrimp remobilized, on average, ~300 g dry weight (DW) of sediment d^{-1} (range, wet to dry months = 204 to 346 g d^{-1}), which projects to 112 kg y^{-1}, and harvested 0.8 g DW of leaves d^{-1} (range = 0.2 to 1.1 g d^{-1}), or a total of 291.3 g y^{-1}. The estimated sediment-reworking rate falls within a considerable range of 0.8 to 1.4 kg m^{-2} d^{-1} during the wet and dry months, for an average shrimp density of 2 m^{-2}, and, correspondingly, the estimated leaf harvesting rate of 0.4 to 2.3 g m^{-2} d^{-1} represents moderate herbivory, equivalent to 12 to 42% of leaf production, respectively.

Keywords: Alpheidae, bioturbation, burrowing, herbivory

Introduction

Diverse faunal assemblages inhabit the different physical compartments of seagrass meadows for refuge and food (McRoy and Helfferich 1977, Leber 1985, Larkum et al. 1989). These assemblages account for the complex trophic hierarchy in seagrass meadows, together with large transient species like dugongs, green turtles, and waterfowl (Jacobs et al. 1981, Thayer et al. 1984, Lanyon et al. 1989, Preen 1995). Direct herbivory of seagrass leaves forms a significant pathway in this trophic hierarchy (Valentine and Heck 1999), which may have been underestimated in the past (Klumpp et al. 1989, Hemminga et al. 1991, Duarte and Cebrián 1996, Cebrián et al. 1997). The list of herbivores includes animals of broad size ranges and the smaller ones may contribute substantially to total herbivory, e.g., sea urchins *Lytechinus variegatus* (Lamarck) (Heck and Valentine 1995) and snapping shrimps of the genus *Alpheus* (Alpheidae, Caridea) (Stapel and Erftemeijer 1997).

Burrowing is a more prominent activity of these alpheid shrimps than leaf harvesting, and their active bioturbation features showcases of classical 'ecological engineering' (Jones et al. 1994). This bioturbation behavior, well described for the thalassinidean shrimps (Thalassinidea), has been shown to affect plant dynamics (Suchanek 1983, Valentine et al. 1994, Duarte et al. 1997) which, in turn, contributed to either the fragmentation (Townsend and Fonseca 1998) or, conversely, the maintenance of meadows (Duarte et al. 1997). In Philippine meadows, the snapping shrimp *Alpheus macellarius*, Chace, 1988, is responsible for 'crenate mounds' (*sensu* Dworshack and Ott 1993) or sand patches that may outnumber the sand volcanoes/ mounds of thalassinidean shrimps (Nacorda et al., Chapter 2, this Thesis), and for clipped shoots within the vicinity of the sand patches. Burrowing and leaf harvesting make alpheid shrimps potentially important bioturbation agents as well as grazers, but their significance remains uncertain because a quantification of their behavior so far is lacking. In this paper, we describe the aboveground behavior of *A. macellarius*, and derive estimates of its importance for sediment reworking and leaf herbivory in the relatively pristine and well-studied seagrass meadow off Bolinao (NW Philippines).

Methods

Behavioral observations were carried out in the shallow and pristine meadow of mixed seagrasses located in Silaki off Santiago Island, Bolinao (NW Philippines; 16°26.42 N, 119°55.72 E; Fig. 1) during the El Niño year of 1998 (June-July, September-October), the following year's dry months (April-May 1999) and in October 1999 (wet). The meadow is part of the 27 km^2 of seagrass vegetation that

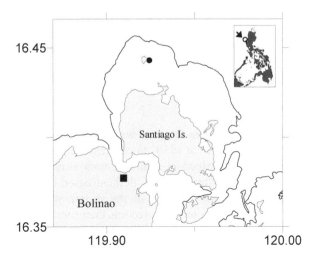

Figure 1. Location of the study site, the seagrass meadow in Silaki (•) within the reef flat
around Santiago Island, relative to that of the marine lab (■) in Bolinao (NW Philippines).

covers 84% of the reef flat surrounding Santiago Island (McManus et al. 1995), has
clear overlying water, is exposed to frequent wave disturbance, and its seagrass
community of 9 species is dominated by *Thalassia hemprichii* (Ehrenberg)
Ascherson (Rollon and Fortes 1991, Vermaat et al. 1995). Sand patches and
mounds (volcanoes) are frequent within this dense meadow, and, on average,
measured 30% (maximum = 52%) and 8% (maximum = 32%) of the bed,
respectively (Nacorda et al., Chapter 2, this Thesis).

The snapping shrimp *Alpheus macellarius* in Silaki occurs with the blue-
speckled prawn goby *Cryptocentrus octafasciatus* Regan, 1908. The shrimp's body
is typically dark green and its first abdominal somite is tinged orange. It carries its
major cheliped extended, like most other alpheid shrimps, and the chela's dactyl and
plunger tips also have the orange coloration. Chace (1988) has placed *A.
macellarius* as a member of the Brevirostris species group, which stands out from
other species by the characteristics of its major and minor chela. The goby, on the
other hand, approximates the color of the sand substrate, and the dark longitudinal
bands and spots along its body help to discern its presence at the openings easily.

One day prior to video recording, a buoy marker was fixed in the middle of sand
patches, then all openings with shrimps ~4 cm in length were selected within 1-m
radius from the buoy and then marked by labeled wooden sticks. A stainless steel
tripod was installed in place of the buoy upon return, in which an underwater
videocamera system was secured. Observations of aboveground activity began at
0700H and proceeded until sundown (~1900H). Video recordings followed a focal
sampling strategy, i.e., activity in a single opening was instantaneously recorded for
5 minutes from the shrimp's first hour of emergence and every hour thereafter until
neither shrimp nor goby guarding activity was observed to occur. Up to four
openings were observed successively every hour. During interim recordings,
sediment dumped by *A. macellarius* was collected on a low-form plastic tray (Fig. 2)
installed at the burrow opening. Scoops from 10 dumping bouts were accumulated

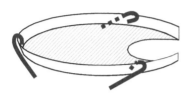

Figure 2. Low-form plastic tray used to collect sediment dumps of *A. macellarius*. The tray was a cutout bottom from a plastic bucket (diameter = 15 cm; height = 0.5 cm) and customized for installation at burrow openings using soft metal pegs as fasteners.

from every shrimp (n = 30) before the heap was collected into plastic bags. In the laboratory, these were oven-dried at 105°C to constant weight.

Shrimp behavior was described in terms of both structure and consequences (Martin and Bateson 1996). Structure was defined based on exclusive behavioral states (excluding the interaction behavior of the shrimp with its goby) identified from a preview of randomly selected videorecords then summarized in an ethogram. The sequence of states was transcribed using an electronic event recorder, and afterwards stored as ASCII data files in a PC utilizing a data transfer protocol (Stamhuis et al. 1996). Data transfers were performed after every 20 minutes of transcribing to allow for resets of the event recorder timer (to prevent memory overloading) and for observer rests (to prevent drifts due to fatigue).

Sequence records were subjected to zero-order Markov-chain analysis, which directly provided frequency and duration summaries for single behavioral states (Stamhuis et al. 1996). Characteristic single-state sequences were afterwards manually assigned into behavioral classes (= consequences) or specific activities, of which frequency and duration summaries were calculated. Values were all expressed as rates, i.e., numbers or time (in seconds) per hour. We pooled our 1998 (ENSO) dataset under one 'dry' period *a posteriori*, since our samplings were carried out during the climate window without the expected intermittent rains. Activity patterns within the observation hours were detected from plots of frequency and duration (*y*-axis) as a function of time (0800-1900H).

We applied repeated measures ANOVA to examine the effect of sampling period (between-subjects factor) on activity frequency and duration variables, which were obtained from the same shrimp individual every hour (within-subjects factor). We included data only for the time window common to all three sampling periods, i.e., between 1000 and 1600H. Results from the univariate approach of the within-subjects tests were used when the data complied with the test's sphericity assumption (Mauchly's $W > 0.05$), otherwise, results from the multivariate approach were utilized (Pillai's Trace statistic) (*sensu* Potvin et al. 1990). Variables with significant effects were subsequently compared using simple contrasts (between-subjects) or Helmert contrasts (within-subjects, within- x between-subjects interaction). Finally, the significance of sediment moving and harvesting behavior was assessed from estimated rates of turnover and harvest biomass during the dry and wet months. The unit leaf biomass harvested was assumed as 50% of mean leaf weight, i.e., ~19 mg DW (Vermaat et al. 1995), and we incorporated a success rate of 75% in the calculations to account for the times the shrimps may have failed to secure their harvest into the burrow. Annual estimates were thereafter derived.

Results

Alpheus macellarius was active during most of the observation period, emerging from its burrow at ~0900H and disappearing from view by ~1900H at the latest. Its goby partner, *Cryptocentrus octafasciatus*, emerged from the burrow earlier (~0700H) and was observed to extend its presence at the opening until past 1900H on quiet full moons. The shrimp's ethogram consisted of 12 exclusive behavior states (Table 1, A) and typical successive states were categorized into four clearly identifiable behavioral sequences or activities (Table 1, B).

The shrimp's aboveground bouts were all very short (bout length, Table 2) and the time interval between any of these bouts was 38 s ($\pm$ 4 SEM), during which the shrimp was out-of-view (i.e., concealed in its burrow, accounting for 90$\pm$1% of its active period at daytime). The time allocated by *A. macellarius* for its aboveground

Table 1. Behavioral states (A) and typical behavioral activities (B) of *Alpheus macellarius* observed in the seagrass meadow off Silaki, Bolinao (NW Philippines).

Category	Description
A. States	
1. Bulldozer	Shovel and lift sediment with both major and minor chela
2. Carry	Walk with chela-load full of sediment; Walk with rubble or fragment clamped by minor chela
3. Dump	Drop carried sediment load at opening with slight push from the major chela
4. Drop	Let go of clamped rubble/ fragment at pile opposite the sediment dump site
5. Touch	Contact of antennae/ chela with seagrass or algae
6. Clamp	Hold rubble/ fragment/ leaf with minor chela
7. Pull	Drag clamped algal or other fragment into the burrow
8. Cut	Cut the clamped leaf/ fragment with plunger of major chela
9. Eat	Pick-up particles from sediments using pereiopods then bring to mouth one after another
10. Walk	Move forward on sediment without carrying anything on chela
11. Pause	Brief stop- body still but with antennal movements
12. Retreat	Walk backward on sediment without carrying anything on chela; Return to burrow tail first, e.g., after dumping sediment or cutting leaf/fragment; Disappear suddenly from view due to disturbance, e.g., approaching fish
B. Activities	
1. Move sediment	Sequences of carrying sediment and dumping, sometimes with further bulldozing in between (carry-[bulldozer]-dump-retreat)
2. Stack rubble	Sequences of carrying and dropping a piece of rubble or shell fragment, sometimes moving/transferring dropped material to a stable spot ([clamp]-carry-drop-retreat)
3. Survey	Sequences of walking and stopping at the opening (walk-pause), and sometimes approaching and 'feeling' seagrass then returning to and stopping at the opening (walk-touch-retreat-pause)
4. Harvest	Sequences of walking, touching, clamping, and cutting leaf or algae, and in some instances, clamped algae were merely pulled down into the burrow (walk-touch-clamp-[pull]-cut-retreat); the sequence 'walk-pause-eat-retreat' was subsumed under this functional class

Table 2. Frequencies, bout lengths, durations (both in seconds), and time allocations (% of total active period) of the activities of *A. macellarius* in the seagrass meadow off Silaki (Bolinao, NW Philippines). Frequency and duration values are average rates per hour ($\pm$ SEM) pooled for 1 day in each sampling period. Superscripts denote significant 'seasonal' differences on frequency (lowercase) and duration (uppercase) data (simple contrasts, $p < 0.05$, following repeated measures ANOVA, Table 3).

Sampling period	Variable	Activity Move sediment		Stack rubble		Survey		Harvest		(Hidden)	
Dry 1998	Frequency	72	± 13	24[b]	± 8	22[b]	± 6	10	± 6	129	± 22
	Duration	196	± 45	78[B]	± 20	155[B]	± 42	59	± 18	3106[A]	± 102
	Bout length	2.6	± 0.3	3.9	± 0.5	7.2	± 0.7	10.6	± 3.7	29.5	± 6.4
	% of time	5.5	± 1.3	2.2	± 0.6	4.3	± 1.2	1.7	± 0.5	86.4	± 2.7
Dry 1999	Frequency	81	± 10	12[ab]	± 3	21b	± 3	8	± 1	108	± 11
	Duration	208	± 40	38[A]	± 11	104[AB]	± 23	52	± 13	3338[AB]	± 58
	Bout length	2.5	± 0.2	2.8	± 0.4	4.9	± 0.4	6.1	± 0.8	33.5	± 2
	% of time	5.3	± 1.1	1	± 0.3	2.6	± 0.5	1.3	± 0.4	89.9	± 1.6
Wet 1999	Frequency	51	± 12	7[a]	± 2	10a	± 2	2	± 1	72	± 11
	Duration	119	± 26	26[A]	± 10	46[A]	± 16	69	± 50	336[AB]	± 42
	Bout length	2.6	± 0.4	3.8	± 0.8	4.6	± 0.4	17	± 8.5	53.5	± 9.9
	% of time	3.2	± 0.8	0.7	± 0.2	1.3	± 0.4	1.6	± 1.1	93.2	± 4
Pooled	**Frequency**	**70**	**± 7**	**14**	**± 3**	**18**	**± 2**	**7**	**± 2**	**102**	**± 9**
	Duration	**179**	**± 23**	**46**	**± 9**	**103**	**± 18**	**59**	**± 15**	**3279**	**± 45**
	Bout length	**2.6**	**± 0.2**	**3.4**	**± 0.3**	**5.5**	**± 0.4**	**10.5**	**± 2.7**	**38.1**	**± 4**
	% of time	**4.8**	**± 0.6**	**1.2**	**± 0.2**	**2.7**	**± 0.5**	**1.5**	**± 0.3**	**89.8**	**± 1.2**

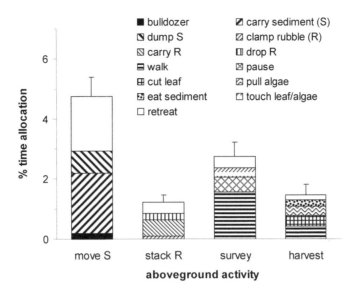

Figure 3. Aboveground time budget of *A. macellarius* in the field expressed as percentages of time per activity (+ SEM) and showing the contribution of the behavioral states included in each activity (n = 21 shrimps).

functions was altogether only 10% of its total active time (Table 2): moving sediment was displayed the most, followed by surveying, harvesting, and finally, stacking up rubble (Fig. 3). Carrying and retreat bouts accounted for >75% of the time for moving sediments and stacking up rubble, surveying mostly involved slow walks to the opening, and harvesting was mostly described by walking and subsequent cutting and pulling bouts.

All four behavioral sequences were regularly displayed by *A. macellarius* (Fig. 4). In 1998, activity commenced at 0900H and proceeded for 9 hours, and frequencies tended to increase towards midday (Fig. 4, top graphs). Patterns were not as apparent in 1999 (Fig. 4, middle and bottom graphs). Shrimp activity during the dry months started 1 h later (1000H) than previously recorded but still continued for 9 hours. Frequencies did not significantly differ from the previous sampling ($p >$ 0.05, Tables 2, 3A), hence, pooled means may be utilized to represent activity on typical dry months: move sediment, 77 (± 8) h^{-1}; stack rubble, 17 (± 4); survey, 21 (± 3); and harvest, 9 (± 2). On the following wet period, shrimp activity was observed for 8 hours and frequencies were reduced by 34% for moving sediment, 52% for surveying, 78% for harvesting, and 59% for stacking up rubble (Table 2). The decrease in the frequencies for stacking and surveying activities was significant ($p <$ 0.05, Tables 2, 3A).

Total duration of each aboveground activity followed the trends for frequencies (Fig. 4, Tables 2, 3B). Pooled daily allocations reach up to only 10 (± 2) and 7 (± 1)% during the dry and wet periods, respectively, or a range of ~2 to as much as nearly 30% on the dry months (Fig. 5) and only close to 10% on the wet month of 1999. On the other hand, the time spent by the shrimp in the burrow ('hidden') significantly increased during the wet period (1999), and was quite variable among observation hours ($p <$ 0.05, Tables 2, 3B).

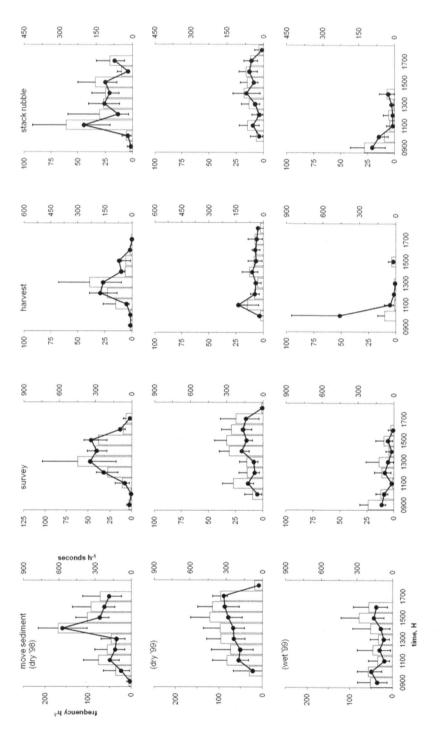

Figure 4. Frequencies (bars + SEM) and duration (filled circles ± SEM) of aboveground activities by *A. macellarius* from 0900 to 1800H during three sampling periods in the Silaki meadow (Bolinao, NW Philippines). No shrimp activity was observed between 0800 to 0859H and, hence, was omitted from the x-axis.

Table 3. Values of *F* or approximate *F* ($^+$, converted from Pillai's Trace statistic) from
repeated measures ANOVA, which examined the effect of sampling period on the hourly
frequency (A) and duration (B) of *A. macellarius* activities. Asterisks (*) denote
significant effects ($\alpha = 0.05$); differences among sampling periods are given in Table 2.
Legend: [a], wet '99 > dry '98 (1100, 1400, and 1500H datasets) and dry '99 > dry '98
(1400 and 1500H).

Variable / Factor	df	Activity				
		Move sediment	Stack rubble	Survey	Harvest	Hidden
A. Frequency						
Between-subjects test						
Sampling period	2	2.043	4.604 *	4.715 *	1.669	-
Within cells (error)	18					
Within-subjects test						
Hourly frequency	6	1.331	0.716 $^+$	2.205 $^+$	0.521 $^+$	-
Hourly frequency x Sampling period	12	0.502	1.171 $^+$	1.653 $^+$	0.897 $^+$	-
Within cells	108					
B. Duration						
Between-subjects test						
Sampling period	2	1.886	5.475 *	5.116 *	0.135	4.600 *
Within cells	18					
Within-subjects test						
Hourly duration	6	1.603	1.007	1.662	0.680 $^+$	2.157 $^+$
Hourly duration x Sampling period	12	1.117	1.322	1.502	0.751 $^+$	2.436 $^{+* a}$
Within cells	108					

A single full scoop of sediment carried and subsequently dumped by *A. macellarius* was 0.50 ± 0.03 g in dry weight. Results of subsequent calculation indicated weights of between 204 and 346 g of sediment moved per day, which, when projected over an area of 1 m^2 with 2 active shrimps, translates to between 0.8 and 1.4 kg of reworked sediments during the wet and dry months, respectively (Table 4A).

The shrimp harvested leaf fragments or full-length leaves of small seagrasses regardless of the species, and commonly nipped these at just above the substratum level in *Syringodium isoetifolium* (Ascherson) Dandy, past the ligula in *T. hemprichii*, past the sheath in *Cymodocea rotundata* Ehrenberg *et* Hemprich *ex* Ascherson, *C. serrulata* (R. Brown) Ascherson, and *Halodule uninervis* (Forsskål) Ascherson, or above the spathe in *Halophila ovalis* (R. Brown) Hooker *f*. Harvested leaf fragments ranged considerably in size because seagrass leaves elongate with age until maturity, are produced continuously, and are subject to sloughing and partial herbivory. Resulting estimates of specific harvest rates range, therefore, between 0.2 (wet months) and 1.1 g d^{-1} shrimp^{-1} (dry months) or between 0.4 and 2.3 g m^{-2} d^{-1}

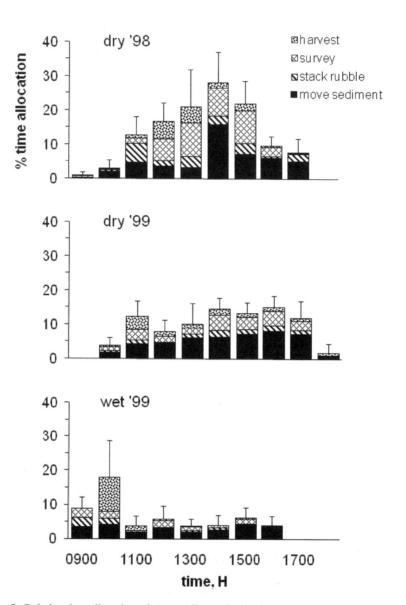

Figure 5. Relative time allocation of *A. macellarius* for its aboveground activities from 0900 to 1900H on three sampling occasions in the clear-water seagrass meadow of Silaki, Bolinao (NW Philippines). Error bars are SEMs of the pooled activity totals.

(Table 4B). The latter range represents a relocation of 8 to 42% of leaf production to the burrows.

Generally, after all living fronds within its 'working radius' were harvested, *A. macellarius* would open new burrows close to its patch border under a canopy of shoots. Thereafter, we assume it to resume its dumping and harvesting bouts at similar rates. Hence, within a year, individual shrimps are projected to have reworked 112 kg of sediments and relocated 291.3 g of leaf material to their burrows.

Table 4. Estimates of the potential impact of sediment movement and leaf harvesting by *A. macellarius* in the seagrass meadow of Silaki (Bolinao, NW Philippines).

	Wet months	Dry months	Annual
A. Sediment movement			
Frequency, h^{-1}	51[a]	77[b]	70[a]
Number of hours aboveground per day	8	9	8.5
Daily dumping rate, d^{-1}	408	693	595
Σ weight dumped per shrimp [c]	204	346	298
(= specific reworking rate, g d^{-1})			
Number of days	137 [d]	228 [e]	
Estimated bulk of reworked sediment, kg [f]	27.9	79.0	112 [g]
Sediment reworking rate, kg m^{-2} d^{-1} [h]	0.8	1.4	1.2
B. Leaf harvesting			
Frequency (h^{-1})	2[a]	9[b]	7[a]
Frequency of successful harvests [i]	1.5	6.8	5.2
Daily harvest rate (d^{-1})	12	60.8	44.6
Σ biomass harvested per shrimp [j]	0.2	1.1	0.8
(= specific harvest rate, g d^{-1})			
Estimated total biomass of harvested leaves, g [f]	30.9	260.4	291.3 [g]
Leaf harvesting rate, g m^{-2} d^{-1} [h]	0.45	2.3	1.7
Relocated production [k]	8%	42%	31%

Notes:
[a] from Table 2
[b] pooled frequency for dry months
[c] dry weight of single dump = 0.50 ($\pm$ 0.03) g
[d] days with rainshowers, tropical depressions, storms, and typhoons between May and October in Iba, (Zambales), Vigan (Ilocos Sur), and Dagupan (Pangasinan) – climate stations facing the South China Sea that were assumed to mirror the conditions in Bolinao; value represents the average for 1998 and 1999 (source: PAGASA, Quezon City, Philippines)
[e] dry (no-rain) days between November and April in the 3 climate stations above, averaged for 1998 and 1999 (source: PAGASA, Quezon City, Philippines)
[f] specific rate multiplied by number of days in column (then converted as appropriate); this estimate is for 1 shrimp individual
[g] sum of values from wet and dry months
[h] specific rate multiplied by 2 ($\pm$0.2) active shrimp individuals per m^2
[i] 75% of frequency per hour
[j] daily harvest rate multiplied by 18.8 mg – represents 50% of mean leaf weight of all small seagrass species in the meadow for each successful harvest; leaf weights obtained from Vermaat et al. (1995)
[k] expressed as % of mixed bed daily leaf productivity, i.e., 5.4 g m^{-2} d^{-1}; extracted from Vermaat et al. (1995)

Discussion

Our observations across several seasons show that this alpheid is mainly engaged with burrowing/mining, and spends most of its time underground. The daily activity rhythm of *Cryptocentrus octafasciatus* and of *Alpheus macellarius* appears similar to the pattern reviewed by Karplus (1987) for other goby-shrimp associations. *Alpheus macellarius* allocated substantial time in maintaining the fragile structures of the burrow by its dumping and stacking bouts. Maintenance bouts also, consequently, present an assurance of the shrimp's continued access to target harvests at the sediment surface. For *C. octafasciatus* which is an omnivore (Froese and Pauly 2001) and which does not seem to actively participate in excavation bouts ('endoecism'; Karplus 1987, Atkinson and Taylor 1991), dumping by *A. macellarius* delivers sediment with particulate organics which are usually filtered through their gills following a series of mouth grabs (Karplus 1987, pers. obs.).

The burrowing behavior of the presently studied alpheid causes substantial sediment reworking. Our estimate range of between 0.8 and 1.4 kg DW m^{-2} d^{-1} is modest compared to reworking rates reported for other thalassinidean shrimps in both tropical and temperate waters (from 0.8 to 12.1 kg m^{-2} d^{-1}; Roberts et al. 1981, Suchanek 1983, Vaugelas 1985, Branch and Pringle 1987, Rowden et al. 1998, Berkenbusch and Rowden 1999) mainly due to differences in shrimp density. Further extrapolation from the quantities of sediment expelled to volume (density of wet sediment = 2.44 g/ml; burrow opening diameter = 4 cm; sediment water content = 34%, Kamp-Nielsen et al. 2002) leads to burrow lengths of 27 to 45 cm extended per day by an individual shrimp within an established burrow. These extensions exceed the tunnel length increases covered by *Callianassa subterranea* (3 cm d^{-1}; Stamhuis et al. 1997) or by burrowing terrestrial invertebrates, e.g., *Lumbricus terrestris* or *Aporretodea longa* (2 cm d^{-1}; Ligthart and Peek 1997). A significant part of the burrow may be ventilated substantially by the shrimp's intermittent pumping behavior, e.g., as observed for *Alpheus mackayi* Banner and Banner (Gust and Harrison 1981). Ventilated burrows expand the sediment's oxic zone (Andersen and Kristensen 1991, Frouin 2000) and may help counteract hypoxia in seagrass rhizomes and roots during periods of light reduction (*sensu* Hemminga 1998).

Alpheus macellarius harvested fresh seagrass leaves and never foraged on these materials at the sediment surface. Instead, the shrimp was observed to hoard the harvest belowground. This behavior seems to qualify for the 'security hypothesis' (Vander Wall 1990), which states that foraging areas (here the seagrass meadow) are less secure and that animals maximize security by returning to a refuge (i.e., burrows) to eat food items they have gathered. Our observations in an accompanying laboratory study revealed that *A. macellarius* usually inserted the leaves to the roof of its burrow then proceeded with its dumping or stacking bouts; while in the burrow, it consumed the stored fronds from the margins. We did not observe harvesting of leaf litter or sheaths, as has been documented for *A. edamensis* De Man, 1988 (Stapel and Erftemeijer 1997). There were, however, instances of particle feeding at the sediment surface, suggesting a preference for the smaller, organically-rich particulates over larger detrital items, akin to other deposit feeders (Taghon 1982, Karplus 1987). Overall, as a leaf hoarder, grazer, and deposit-feeder, *A. macellarius* does not appear to be resource-limited in the meadow.

The observed seasonality in shrimp activities is remarkable, and we speculate on its correlation with shifting monsoons. In addition, leaf harvesting varied substantially among seasons. Lower removal rates occurred during the wet months and coincided with the period of lower seagrass productivity (Vermaat et al. 1995, Agawin et al. 1997). At the shrimp's harvesting rate, our estimate of relocated production of 8 to 42% overlaps with and well exceeds the herbivory attributed to the rabbitfish *Siganus fuscescens* (5 to 10%; Hernandez et al. 1990, Salita-Espinosa 1992) and may be as important as the grazing potential predicted for the sea urchin *Tripneustes gratilla* (24%; Klumpp et al. 1993) in the area. Harvesting per se, however, may not be detrimental to seagrasses: for example, partial defoliation was shown to enhance growth rates in defoliated shoots of *Thalassia testudinum* Banks *ex* König (Tomasko and Dawes 1989) and *T. hemprichii* (Chapter 5, this Thesis). Reallocation of carbohydrate reserves from the roots and rhizomes to the leaves may be induced (Hemminga 1998), hence, leaf growth rates may not be necessarily depressed (Cebrián et al. 1998). The potential food limitation by leaf availability can only be assessed on a larger bed- and longer time-scale, because the gaps, like territories, shifted positions with time (Bell et al. 1999, pers. obs.) as a result of simultaneous shrimp activity and seagrass recolonization. The gaps generated by the shrimps averaged 0.30 m^2 ($\pm$ 0.03), which, similar to thalassinid mounds, may be rapidly recolonized by smaller seagrass pioneers of the meadow through clonal growth (Duarte et al. 1997). These pioneers, hence, may initiate a succession sequence, which is viewed as one of the mechanisms in the maintenance of mixed meadows (Hemminga and Duarte 2000.). In addition, the transfer of leaf material to the burrows represents a process of nutrient conservation within the meadow, and must be included among the inputs that maintain mixed meadows (Hemminga et al. 1991).

Acknowledgments. This study was part of Project WB84-413 funded by WOTRO. We thank Boyet Elefante[†] for fabricating the stainless steel tripod, Angelo Pernetta, Ronald Gijlstra, and José Vós for assistance in sediment sampling and video recording sessions, and Jack Rengel, Criz Ragos, and Macoy Ponce for logistic support during field visits. Dr. Charles Fransen identified the alpheid shrimps and Nadia Palomar provided the taxonomic identity of the goby. HEN is grateful for the advice extended by Dr. Anette Juinio-Meñez during various stages of the study, and to the constructive comments of Prof. Anthony Underwood and two anonymous referees.

References

Agawin, N.S.R., Duarte, C.M., Fortes, M.D., Uri, J.S., and Vermaat, J.E. 2001. Temporal changes in the abundance, leaf growth and photosynthesis of three co-occurring Philippine seagrasses. Journal of Experimental Marine Biology and Ecology 260: 217-239.

Andersen, F.O. and Kristensen, E. 1991. Effects of burrowing macrofauna on organic matter decomposition in coastal marine sediments. In: Meadows, P.S. and Meadows, A., eds., The environmental impact of burrowing animals and animal burrows. The Proceedings of a Symposium held at the Zoological Society of London, 3-4 May 1990. Symposium of the Zoological Society of London 63: 69-88.

Atkinson, R.J.A. and Taylor, A.C. 1991. Burrows and burrowing behaviour of fish. In: Meadows, P.S. and Meadows, A., eds., The environmental impact of burrowing animals and animal burrows. The Proceedings of a Symposium held at the Zoological Society of London, 3-4 May 1990. Symposium of the Zoological Society of London 63: 133-156.

Bell, S.S., Robbins, B.D., and Jensen, S.L. 1999. Gap dynamics in a seagrass landscape. Ecosystems 2: 493-504.

Berkenbusch, K. and Rowden, A.A. 1999. Factors influencing sediment turnover by the burrowing ghost shrimp *Callianassa filholi* (Decapoda: Thalassinidea). Journal of Experimental Marine Biology and Ecology 238: 283-292.

Branch, G.M. and Pringle, A. 1987. The impact of the sand prawn *Callianassa kraussi* Stebbing on sediment turnover on bacteria, meiofauna, and benthic microflora. Journal of Experimental Marine Biology and Ecology 107: 219-235.

Cebrián, J., Duarte, C.M., Agawin, N.S.R., and Merino, M. 1998. Leaf growth response to simulated herbivory: a comparison among seagrass species. Journal of Experimental Marine Biology and Ecology 220: 67-81.

Cebrián, J., Duarte, C.M., Marbá, N., and Enriquez, S. 1997. Magnitude and fate of the production of four co-occurring Western Mediterranean seagrass species. Marine Ecology Progress Series 155: 29-44.

Chace, F.A. Jr. 1988. The caridean shrimps (Crustacea: Decapoda) of the Albatross Philippine Expedition, 1907-1910, Part 5: Family Alpheidae. Smithsonian Contributions to Zoology No. 466. Smithsonian Institution Press, Washington, D.C. total number of p.

Duarte, C.M. and Cebrián, J. 1996. The fate of marine autotrophic production. Limnology and Oceanography 41: 1758-1766.

Duarte, C.M., Terrados, J., Agawin, N.S.R., Fortes, M.D., Bach, S., and Kenworthy, W.J. 1997. Response of a mixed Philippine seagrass meadow to experimental burial. Marine Ecology Progress Series 147: 285-294.

Dworschak, P.C. and Ott, J.A. 1993. Decapod burrows in mangrove-channel and back-reef environments at the Atlantic Barrier Reef, Belize. Ichnos 2: 277-290.

Froese, R. and Pauly, D.: FishBase, World Wide Web electronic publication. www.fishbase.org, 01 Dec 2001.

Frouin, P. 2000. Effects of anthropogenic disturbances of tropical soft bottom communities. Marine Ecology Progress Series 194: 39-53.

Gust, G. and Harrison, J.T. 1981. Biological pumps at the sediment-water interface: mechanistic evaluation of the alpheid shrimp *Alpheus mackayi* and its irrigation pattern. Marine Biology 64: 71-78.

Heck, K.L. Jr. and Valentine, J.F. 1995. Sea urchin herbivory: evidence for long-lasting effects in subtropical seagrass meadows. Journal of Experimental Marine Biology and Ecology 189: 205-217.

Hemminga, M.A. 1998. The root/ rhizome system of seagrasses: an asset and a burden. Journal of Sea Research 39: 183-196.

Hemminga, M.A. and Duarte, C.M. 2000. Seagrass ecology. Cambridge University Press, Cambridge, 298 p.

Hemminga, M.A., Harrison, P.G., and van Lent, F. 1991. The balance of nutrient losses and gains in seagrass meadows. Marine Ecology Progress Series 71: 86-96.

Hernandez, H.B., Aliño, P.M., and Jarre, A. 1990. The daily food consumption of *Siganus fuscescens*, an important herbivore in seagrass communities at Bolinao, Pangasinan province, Philippines. ICES C.M. 16, G:18, 15 p.

Jacobs, R.P.W.M., den Hartog, C., Braster, B.F., and Carriere, F.C. 1981. Grazing of the seagrass *Zostera noltii* by birds at Terschelling Dutch Wadden Sea. Aquatic Botany 10: 241-259.

Jones, C.G., Lawton, J.H., and Schack, M. 1994. Organisms as ecosystems engineers. Oikos 69: 373-386.

Kamp-Nielsen, L., Vermaat, J.E., Wesseling, I., Borum, J., and Geertz-Hansen, O. 2002. Sediment properties along gradients of siltation in South-East Asia. Estuarine, Coastal and Shelf Science 54: 127-137.

Karplus, I. 1987. The association between gobiid fishes and burrowing alpheid shrimps. Oceanography and Marine Biology Annual Reviews 25: 507-562.

Klumpp, D.W., Howard, R.K., and Pollard, D.A. 1989. Trophodynamics and nutritional ecology of seagrass communities. In: Larkum, A.W.D., McComb, A.J., and Shepherd, S.A., eds., Biology of seagrasses, a treatise on the biology of seagrasses with special reference to the Australian region, pp. 394-457. Elsevier Science Publishers B.V., Amsterdam.

Klumpp, D.W., Salita-Espinosa, J.T., and Fortes, M.D. 1993. Feeding ecology and trophic role of sea urchins in a tropical seagrass community. Aquatic Botany 45: 205-229.

Lanyon, J., Limpus, C.J., and Marsh, H. 1989. Dugongs and turtles: grazers in the seagrass system. In: Larkum, A.W.D., McComb, A.J., and Shepherd, S.A., eds., Biology of seagrasses, a treatise on the biology of seagrasses with special reference to the Australian region, pp. 610-634. Elsevier Science Publishers B.V., Amsterdam.

Larkum, A.W.D., McComb, A.J., and Shepherd, S.A., eds. 1989. Biology of seagrasses, a treatise on the biology of seagrasses with special reference to the Australian region. Elsevier Science Publishers B.V., Amsterdam, 814 p.

Leber, K.M. 1985. The influence of predatory decapods, refuge, and microhabitat selection on seagrass communities. Ecology 66: 1951-1964.

Ligthart, T.N. and Peek, G.J.C.W. 1997. Evolution of earthworm burrow systems after inoculation of lumbricid earthworms in a pasture in the Netherlands. Soil Biology and Biochemistry 29: 453-462.

Martin, P. and Bateson, P. 1996. Measuring behavior, an introductory guide. Cambridge University Press, Cambridge, 200 p.

McManus, J.W., Nañola, C.L., Reyes, R.B., and Kesner, K.N. 1995. The Bolinao coral reef resource system. In: Juinio-Meñez, M.A. and Newkirk, G.F., eds., Philippine coastal resources under stress – selected papers from the 4th Annual Common Property Conference, Manila, June 1993, pp. 193-204. Coastal Resources Research Network, Dalhousie University, Nova Scotia, Canada and Marine Science Institute, University of the Philippines, Quezon City.

McRoy, C.P. and Helfferich, C., eds. 1977. Seagrass ecosystems, a scientific perspective. Marcel Dekker Inc, New York, USA, 314 p.

Nacorda, H.M.E., Cayabyab, N.M., Fortes, M.D., and Vermaat, J.E. 2008. Chapter 2: The distribution of burrowing shrimp disturbance in Philippine seagrass meadows. (this Thesis)

Potvin, C., Lechowicz, M.J., and Tardiff, S. 1990. The statistical analysis of ecophysiological response curves obtained from experiments involving repeated measures. Ecology 71: 1389-1400.

Preen, A. 1995. Impacts of dugong foraging on seagrass habitats: observational and experimental evidence for cultivation grazing. Marine Ecology Progress Series 124: 201-213.

Roberts, H.H., Wiseman W.J. Jr., and Suchanek, T.H. 1981. Lagoon sediment transport: the significant effect of *Callianassa* bioturbation. Proceedings of the 4th International Coral Reef Symposium, Manila 1: 459-465.

Rollon, R.N. and Fortes, M.D. 1991. Structural affinities of seagrass communities in the Philippines. In: Alcala, A.C., ed., Proceedings of the Regional Symposium on Living Resources in Coastal Areas, 30 Jan-1 Feb 1989, Manila, pp. 333-346. Marine Science Institute, University of the Philippines, Quezon City.

Rowden, A.A., Jones, M.B., and Morris, A.W. 1998. The role of *Callianassa subterranea* (Montagu) (Thalassinidea) in sediment resuspension in the North Sea. Continental Shelf Research 18: 1365-1380.

Salita-Espinosa, J.T. 1992. Aspects of the feeding biology of juveniles of the white-spotted spinefoot, *Siganus fiscescens* Pisces: Siganidae. M.Sc. Thesis, Marine Science Institute, University of the Philippines, Quezon City, 75 p.

Stamhuis, E.J., Reede-Dekker, T., van Etten, Y., de Wiljes J.J., and Videler, J.J. 1997. Behavior and time allocation of the burrowing shrimp *Callianassa subterranea* (Decapoda, Thalassinidea). Journal of Experimental Marine Biology and Ecology 204: 225-239.

Stamhuis, E.J., Schreurs, C.E., and Videler, J.J. 1997. Burrow architecture and turbative activity of the thalassinid shrimp *Callianassa subterranea* from the central North Sea. Marine Ecology Progress Series 151: 155-163.

Stapel, J. and Erftemeijer, P.L.A. 1997. Leaf harvesting and sediment reworking by burrowing alpheid shrimps in a *Thalassia hemprichii* meadow in South Sulawesi, Indonesia. In: Stapel, J., Nutrient dynamics in Indonesian seagrass beds: factors determining conservation and loss of nitrogen and phosphorus, pp. 33-41. PhD Thesis, Katholieke Universiteit Nijmegen, WOTRO/ NWO.

Suchanek, T.H. 1983. Control of seagrass communities and sediment redistribution by *Callianassa* (Crustacea, Thalassinidea) bioturbation. Journal of Marine Research 41: 281-298.

Taghon, G.L. 1982. Optimal foraging by deposit-feeding invertebrates: roles of particle size and organic coating. Oecologia 52: 295-304.

Thayer, G.W., Bjorndal, K.A., Ogden, J.C., Williams, S.L., and Zieman, J.C. 1984. Role of larger herbivores in seagrass communities. Estuaries 7: 351-376.

Tomasko, D.A. and Dawes, C.J. 1989. Effects of partial defoliation on remaining intact leaves in the seagrass *Thalassia testudinum* Banks *ex* König. Botanica Marina 32: 235-240.

Townsend, E.C. and Fonseca, M.S. 1998. Bioturbation as a potential mechanism influencing spatial heterogeneity of North Carolina seagrass beds. Marine Ecology Progress Series 169: 123-132.

Valentine, J.F. and Heck K.L. Jr. 1999. Seagrass herbivory: evidence for the continued grazing of marine grasses. Marine Ecology Progress Series 176: 291-302.

Valentine, J.F., Heck K.L. Jr., Harper, P., and Beck, M. 1994. Effects of bioturbation in controlling turtlegrass *Thalassia testudinum* abundance: evidence from field enclosures and observations in the Northern Gulf of Mexico. Journal of Experimental Marine Biology and Ecology 178: 181-192.

Vander Wall, S.B. 1990. Food hoarding in animals. The University of Chicago Press, Chicago, 445 p.

Vaugelas, J. de: Sediment reworking by callianassid mud-shrimp in tropical lagoons: a review with perspectives. Proceedings of the 5th International Coral Reef Congress, Tahiti, 6: 617-622.

Vermaat, J.E., Agawin, N.S.R., Duarte, C.M., Fortes, M.D., Marbá, N., and Uri, J.S. 1995. Meadow maintenance, growth and productivity of a mixed Philippine seagrass bed. Marine Ecology Progress Series 124: 215-225.

Chapter 4

Burrows and behavior of the snapping shrimp *Alpheus macellarius,* Chace, 1988, in different seagrass substrates

H.M.E. Nacorda[1, 2], E.J. Stamhuis[3], J.E. Vermaat[2, 4]

[1] Marine Science Institute, University of the Philippines, UPPO Box 1, Diliman, Quezon City, 1101 The Philippines

[2] Department of Environmental Resources, UNESCO-IHE Institute for Water Education, PO Box 3015, 2601 DA Delft, The Netherlands

[3] Department of Marine Biology, Rijksuniversiteit Groningen, PO Box 14, 9750 AA Haren, The Netherlands

[4] present address: Institute for Environmental Studies, Vrije Universiteit, De Boelelaan 1087, 1081 HV Amsterdam, The Netherlands

Abstract

Burrowing alpheid shrimps occur commonly in seagrass meadows and as such, are presumed to respond to the various sediment environments in which seagrasses are found. To determine the effects of sediment type on seagrass-associated alpheid shrimps, we examined burrows and the burrowing behavior of *Alpheus macellarius*, Chace, 1988, in three bed substrates – sand, muddy sand, and sandy mud – in laboratory cuvettes. *Alpheus macellarius* immediately excavated into all the substrates at initiation of the experiment and successfully established the first burrows within 2 hours (sand) to 1 day (muddy sand, sandy mud). Burrowing effort was highest and tunnel lengths were most extensive in sand after 26 days ($p < 0.05$). After 37 days, shrimp activity in all substrates was observed to shift to considerable wandering and feeding (mainly particle ingestion) became conspicuous. Burrowing behavior, though reduced, remained significantly higher in sand than in the other two substrates ($p < 0.05$). Overall, soft sandy sediments presented greater support for the burrowing behavior of *A. macellarius*, which, with reinforcement from dense seagrass, probably accounts for higher burrow densities in sand in the field. Conversely, shrimp burrowing appeared limited and substituted by concealment strategies in terrigenous substrates. Hence, with less support from sparser vegetation in the field, burrow numbers were also reduced.

Introduction

Various decapod crustaceans are strongly associated with seagrass beds for shelter, refuge from predators, and food (Leber 1984, Zupo and Nelson 1999, Gallmetzer et al. 2005, Al-Maslamani et al. 2007). Shrimps and crabs, in particular, browse and burrow in sediments, which have known effects on the habitat fabric and on infaunal communities ('ecosystem engineering', *sensu* Jones et al. 1994, Levinton 1995, Berkenbusch et al. 2000, Berkenbusch and Rowden 2003, 2007). The significance of this disturbance in various meadows has been highlighted for the cryptic yet broadly studied thalassinidean shrimps (Thalassinidea) (Frey and Howard 1975, Ott et al. 1976, Suchanek 1983). Features of the burrows of these animals (i.e., the presence of pieces of seagrass or algae in the tiered galleries, chambers, and burrow lining/ walls) and the conspicuous sediment mounds at the sediment surface were seen to provide indications of active sediment processing, which conform primarily to a deposit-feeding mode (Vaugelas 1985, Griffis and Suchanek 1991, Nickell and Atkinson 1995). The burrowing caridean shrimps, notably the Brevirostris group within the genus *Alpheus* (snapping shrimps) associated with gobiid fishes, constructed and maintained shallow burrows (Karplus 1987), which may become complex as gravel content increased (Palomar et al. 2005). These shrimps similarly exhibited a deposit-feeding mode (Karplus 1987, Palomar et al. 2005).

In mixed-seagrass species meadows of the tropics, bioturbation by these snapping shrimps was recently found to have direct negative effects on the growth of seagrasses (Stapel and Erftemeijer 2000). More recent field surveys in the Philippines have indicated that there can be differences in the numbers of burrow openings and bare sand patches of alpheid shrimps across the different sediment conditions covered by seagrass meadows (Nacorda et al., Chapter 2, this Thesis), which implied possible differences in the burrowing effort in different sediments. Seagrass biomass has been shown to decrease in silty sediments (Terrados et al. 1998) so that seagrass densities may also potentially contribute to the numbers of burrow openings and bare sand patches observed. *In situ* observations revealed that the shrimps frequently expelled sediment from their burrows and also harvested fresh seagrass leaves (Nacorda et al., Chapter 3, this Thesis; Palomar et al. 2005). However, the shrimps spent considerable time underground, which presents a gap in our knowledge of their full diel behavioral pattern. Therefore, the significance of much of the shrimp's belowground activities has remained obscure so far.

Here we report on an extension of our previous field observations by determining the burrowing activity and documenting the behavior of *Alpheus macellarius* Chace, 1988, in three common types of seagrass substrata, using narrow-but-high, transparent cuvettes (*sensu* Stamhuis et al. 1996), allowing belowground behavior to be observed. We hypothesized that the range of substrate types suitable for seagrasses produced different effects on the behavior and burrowing success of *A. macellarius*, thus, we aimed to examine the influence of substrate type on shrimp activity pattern since substrate type may well affect the shrimps' energy requirements for burrow construction, the subsequent construction time, the resulting burrow lengths and configurations (Griffis and Chavez 1988), and, eventually, their spatial distribution in the meadows. We also expected that the alpheid shrimps, similar to thalassinideans, would spend their within-burrow periods (i.e., hidden from view in field conditions) mainly on further burrow work – extensions, maintenance – and feeding.

Methods

Experimental set-up, field collections, and sampling

A week before field collections, transparent glass cuvettes (45 cm width x 35 cm height x 2 cm depth; n = 12), were prepared and then completely submerged in seawater aquaria at the hatchery facility of the Bolinao Marine Laboratory (BML). Continuous aeration and flow-through seawater supply were provided to these units.

Soft sediments were collected from three seagrass sites within the siltation gradient of the Bolinao-Anda system (McManus et al. 1992, Terrados et al. 1998, Kamp-Nielsen et al. 2002): Site 1 – Silaki (carbonate sand; substrate type 1), Site 2 – Guiguiwanen (terrigenous muddy sand; substrate type 2), and Site 3 – Dolaoan (terrigenous sandy mud; substrate type 3) (Table 1, Fig. 1). Plant material was removed; the sediment was transferred to the cuvettes (with aeration turned off), and then allowed to settle up to a depth of 20 cm. Substrate types 1, 2, and 3 were arranged in a randomized complete block design (Gomez and Gomez 1984) to rid off bias during subsequent recordings of shrimp behavior. Seedlings (n = 10) of *Thalassia hemprichii* (Ehrenberg) Ascherson were planted in each cuvette. Plant density in the setups was chosen as a compromise of availability and observed natural densities (Vermaat et al. 1995, Rollon 1997). Spilled sediment was removed from each aquarium and then flow-through seawater was supplied until all aquaria were filled with clear water, and then sediment level in each cuvette was traced on clear plastic that was superimposed on the front wall of the aquarium. After tracing,

Table 1. Properties of seagrass bed sediments where the shrimp and goby pairs were obtained (Lucero) and into which the animals were introduced (1-3). Data for Silaki (1) and Dolaoan (3) are mean values (± SEM) from Nacorda et al. (Chapter 1, this Thesis); those for Guiguiwanen (2) were from Nacorda (unpubl.). Legend: TOM = total organic matter; MGS=Folk's mean grain size

Source	Lucero	1 – Silaki	2 – Guiguiwanen	3 – Dolaoan
Sediment type	Sand + rubble	Sand	Muddy sand	Sandy mud
TOM, %	4.9 (0.2)	5.8 (0.3)		5.2 (0.5)
MGS, ϕ	1.04 (0.1)	1.3 (0.03)	2.6	1.3
Wentworth's grain size composition (%)				
Granules	14.0 (1.8)	8.6 (0.6)	0.4	7.3 (4.1)
Very coarse	13.9 (1.1)	20.1 (0.3)	1.4	11.9 (2.0)
Coarse sand	18.3 (1.6)	20.1 (0.3)	3.7	11.8 (2.4)
Medium sand	22.5 (1.3)	14.1 (0.2)	16.8	14.9 (1.0)
Fine sand	18.6 (1.4)	16.5 (0.3)	52.5	36.3 (4.0)
Very fine sand	8.1 (1.3)	10.1 (0.3)	15.5	9.2 (1.5)
Mud	4.6 (0.8)	10.5 (1.0)	9.7	8.6 (1.9)

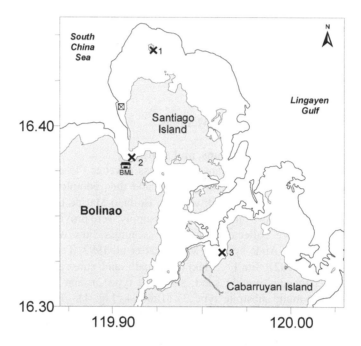

Figure 1. Location of seagrass beds in Bolinao (NW Philippines) where sediment (✕, 1-3) and shrimp and goby pairs (⊠) were collected. The cuvette setups were maintained at the Bolinao Marine Laboratory (BML).

continuous aeration to the setups was resumed and then each aquarium was wrapped on all sides with black canvas and topped with nets after sundown to reduce the presumed influence of artificial light on animal behavior during nighttime and on the overgrowth of algae.

On a separate field visit, 12 naturally co-occurring pairs of the snapping shrimp *Alpheus macellarius* (total length = 4.2 ± 0.1 cm) and the blue-speckled prawn goby *Cryptocentrus octafasciatus* Regan, 1908 (total length = 3.9 ± 0.2 cm) were caught in Lucero (Table 1, Fig. 1), which is a shallow clear-water site with soft sandy sediment. The pairs were caught by inserting a shovel in the heap of sediments immediately outside a burrow opening (without collapsing the entrance), waiting for the animals to resume burrow guarding (goby) and sediment dropping bouts (shrimp), and then trapping the pair by hand in the sediment held by the shovel. These pairs were immediately brought to the hatchery facility, and temporarily held for three hours as separate pairs in containers provided with slow-flowing seawater.

Each alpheid shrimp was carefully introduced with its watchman goby to the cuvette (= Day 0), which was then immediately topped with a gravel-weighted net (mesh = 1 mm) to prevent the animals from escaping and not necessarily impede water flow. Simultaneously, shrimp behavior was recorded following a focal sampling rule (Altmann 1974, Martin and Bateson 1986), using a video camera mounted on a tripod and with the camera's lens maintained at 50 cm distance from the aquarium continuously for 5 minutes. Later, this was reduced to 3 minutes because all sequences had been sufficiently sampled during that period. The development of burrows and changes in sediment levels were monitored after 2 hours (Day 0) and on Days 1, 2, 3, 7, 13, and 26. Configurations were traced on clear plastic (as above) and these traces were subsequently digitized.

The shrimp's activity pattern was determined on Day 37, when burrow lengths remained the same. Our preliminary observations of shrimp feeding indicated that *A. macellarius* demonstrated no preference for any of the small seagrass species, seedlings of *E. acoroides* included. Thus, one day prior to filming, we replenished the lost seedlings (due to burial or consumption by the shrimp) by more readily available shoots (n = 10) of *Cymodocea rotundata* Ehrenberg et Hemprich ex Ascherson. The shrimps' pre-disturbance activities were resumed following a short period of hiding (the state 'retreat') in response to disturbance (Nacorda et al., Chapter 3, this Thesis). Continuous minute video-records were taken for 3 minutes on each shrimp every hour during the day (from 1100 to 1800H) and at nighttime (from 1800 to 2400H; red light was used to facilitate filming).

The maintenance of the setups entailed the weekly removal of algae from the sides of both cuvettes and aquaria and the replacement of top-nets. During cleaning, water levels in the aquaria were ensured to be higher than the sediment level in the cuvettes to prevent the collapse of burrows. Water was subsequently replenished until clear and the flow-through supply was restored.

Data analyses
Shrimp behavior records were transcribed following an 18-state ethogram (Table 2, part A), with each state keyed in on an event recorder during videotape playback (Stamhuis et al. 1996). A type of Markov-chain sequence analysis tested the frequencies of typical behavioral state sequences (from duplets to sextuplets) within each record against the probability of the sequence to occur randomly (Stamhuis et al. 1996). From all significantly occurring sequences, those that were found in at least 5% of all the records for each sediment type were assigned into functional classes (= activities) (Table 2, part B). Frequencies (numbers per hour), bout lengths (seconds), and time allocations (percentage) for these classes were calculated.

A one-way ANOVA was used to examine the effect of substrate type (1, 2, and 3) on bout lengths and time allocations of shrimp activity observed on Day 0. Differences in shrimp behavior parameters on Day 37 (frequency, time allocation, bout length) attributed to the test substrates (between-subjects factor) and to time of day (day, night) (within-subjects factor) were assessed with repeated measures ANOVA. Results from the univariate approach of the within-subjects tests were used when the data complied with the test's sphericity assumption (Mauchly's W > 0.05), otherwise, results from the multivariate approach were utilized (Pillai's Trace statistic; *sensu* Potvin et al. 1990). Simple contrasts were utilized (α = 0.05) to discriminate the substrates where between-subjects effects have been detected ($p <$ 0.05). These statistical analyses were performed in SPSS Release 7.0 (SPSS Inc. 1995).

Burrow construction effort was gauged from tunnel lengths (L_{tunnel}, mm) measured from the sketches/stills, and from volumes of moved sediment ($V_{sediment}$, ml). The latter was calculated as the difference in the measured tunnel areas between two successive monitoring periods then multiplied by the cuvette thickness (= 2 cm). Burrowing velocities (cm tunnel length h^{-1}) were computed from total lengths every monitoring period divided by the number of hours between successive measurements. Cumulative means of total excavated length and the amount of

Table 2. Behavior of *Alpheus macellarius* observed in the laboratory-based cuvettes.

Category	Description
A. States	
1. Bulldozer	Shovel and lift sediment with both major and minor chela
2. Carry	Walk with chela-load full of sediment, or with rubble/ fragment clamped by minor chela, in both cases accompanied by pleopod movements during ascent towards the dump/ drop site
3. Dump/ Drop	Drop carried sediment load at entrance with slight push from the major chela; Pile/ let go of clamped rubble or fragment at pile opposite the sediment dump site
4. Clamp	Hold rubble/ leaf or fragment with minor chela
5. Excavate	Dig sediment with both chela and the first 2 pairs of pereiopods
6. Sweep	Fan pleopods fiercely, sometimes simultaneous with excavating or walking, resulting in loosened and suspended sediments/ particulates
7. Tamp	Fix rubble in place; insert leaf/ fragment into burrow wall or roof
8. Touch	Contact of antennae/ chela with seagrass or goby
9. Pull	Drag clamped rubble/ fragment into the burrow
10. Cut	Cut the clamped leaf/ fragment with plunger of major chela
11. Eat	Hold leaf between both chela then bite from the leaf margin; pick-up grains/ particles with from sediments using pereiopods then bring to mouth one after another; stoop cephalothorax then bite on sediment; or hang and bite on burrow wall
12. Walk	Move forward on sediment without carrying anything on chela
13. Pause	Brief stop; body still but with antennal movements
14. Retreat	Return to burrow tail first, e.g., after dumping sediment or cutting leaf/ fragment, sometimes followed by silt-puffs from the buccal area; walk backward on sediment without carrying anything on chela
15. Turn	Change direction by doing a somersault
16. Ventilate	Fan pleopods slowly while sitting still (pause) or eating
17. Groom	Rub pereiopods; scratch carapace/ pick into cephalothorax, pleopods, or telson using pereiopods
18. Swim	Ascend to swim in the water column, always with pleopod fanning; roam to the glass wall
B. Activities	
1. Burrow	States 'excavate' and 'sweep'; sequences 'excavate-sweep' and 'bulldozer-carry-dump', often with 'pause', 'walk', 'retreat', or 'turn' as initial/ subsumed/ terminal states
2. Survey	States 'touch' and 'tamp'; sequences 'touch-walk', 'touch-walk-pause', 'walk-touch-walk', and 'tamp-walk'
3. Wander	Sequences 'walk-pause', 'retreat-pause', 'turn-walk', and 'turn-pause'
4. Rest	State 'pause' with duration $\geq$ 30 seconds
5. Feed	State 'eat', often with 'retreat', 'pause', 'walk', 'walk-pause', or 'retreat-pause' as initial/ terminal states/ sequences
6. Groom	State 'groom'; sequence 'groom-pause'
7. Ventilate	State 'ventilate'; sequence 'pause-ventilate'

moved sediment for each substrate type were plotted against time (days), then were fitted to the saturation function:

$$y = a \times (1 - e^{-b \times t})$$

with a as the maximum asymptotic length or volume, b, as the increase parameter, and t, time in days (Stamhuis et al. 1997). These parameters were estimated by the iterative least squares fitting procedure in Sigma-Plot v.8 (Jandel Scientific/ SPSS, 2002).

Results

Burrowing behavior and burrow development

Alpheus macellarius immediately executed burrowing bouts in all substrates by swiftly digging in between the *Thalassia* seedlings or at the left/right border of the cuvettes. Burrowing was displayed the most ($69 \pm 4\%$ SEM; bout length = 15.5 ± 4.1 s; Fig. 2, a), and was represented by short duplet and triplet sequences in sand and chains of these sequences in the other substrates. Substrate type did not affect this initial burrowing behavior in terms of either time allocations or bout lengths ($p > 0.05$). Apart from burrowing, the shrimps were also engaged in bouts of wandering ($18 \pm 4\%$; bout length = $7.4 \pm .07$ s) and surveying ($7 \pm 2\%$; bout length = 6.8 ± 4.1 s) or were observed at rest ($6 \pm 3\%$; bout length = 41.3 ± 2.6 s) (Fig. 2, a).

The first burrow openings and short pits were excavated at high velocities either at an angle from or perpendicular to the sediment surface (Fig. 3). These were formed in sand within 2 h after introduction or by Day 2 in muddy sand and sandy mud. Tunnels were extended obliquely from these pits by excavating at much reduced velocities and this continued until new openings were created. There were 1-2 openings, positions of which were often shifted by the shrimps through excavations from the pits rerouted to the surface. Leaf harvesting and storing events

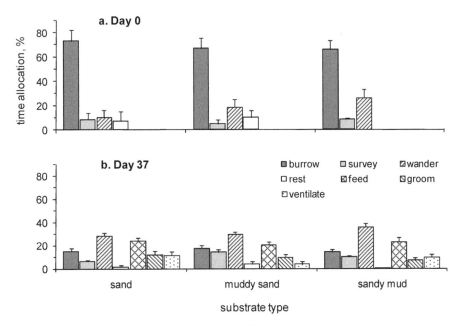

Figure 2. Behavior and time allocation of *A. macellarius* upon introduction to the cuvettes (a) and during activity pattern determinations (b).

were observed directly when tracing the tunnels on transparencies, but did not occur sufficiently frequently to be quantified as behavioral bouts; clipped seedlings were observed to be completely buried already after 7 days.

The constructed burrows were simple and ranged from single J- or U-shapes to interconnected U's in the two-dimensional cuvettes (Fig. 3). Tunnel widths were not regular (minimum ≈ 2.0 cm but accommodated the sizes of the shrimps and gobies) and slight enlargements ('chambers') catered to convenient turns of both animals. In the case of burrow collapse, new segments were immediately made – *A. macellarius* excavated new pits either from the cuvette borders or into grooves, which then were extended as tunnels. After 26 days, excavations in sand produced the longest burrows and reached 54 ± 2 cm; those in muddy sand and sandy mud were shorter (42 ± 7 and 35 ± 1 cm, respectively; Fig. 4, a). The saturation equation explained over 85% of the variation in mean tunnel lengths ($p < 0.01$) and predicted the mentioned maxima (*a*), with $b = 0.44 \pm 0.06$ (sand), 0.10 ± 0.04 (muddy sand), and 0.34 ± 0.03 (sandy mud). These maximum lengths also appeared to have been reached already within 1 week in sand and sandy mud (Fig. 4, a) and within 2 weeks in muddy sand.

Because of intense burrowing during the first few days, net volumes of sand displaced by *A. macellarius* were consequently larger compared with succeeding measurements. In all the substrates, however, shrimps were not only observed to execute bulldozing bouts from the burrows and expel sediments at the surface, but, often also remobilized only surface sediments. The estimated (net) maximum amount of remobilized sand was, hence, 699 ± 33 ml, which exceeded the maxima of either muddy sand (277 ± 6) or sandy mud (195 ± 19) after 26 days (Fig. 4, b). Curve fits similarly explained over 85% of the variation in the remobilized sediment volumes ($p < 0.01$) with $b = 0.34 \pm 0.05$ (sand), 0.68 ± 0.06 (muddy sand), and 0.43 ± 0.14 (sandy mud), and showed maximum tunnel volumes to have been reached within a week (Fig. 4, b).

Shrimp activity pattern

Alpheus macellarius evidently displayed seven activities (= functional classes) in the three test substrates on Day 37: feeding, grooming, and ventilating were added to the four classes displayed on Day 0 (Fig. 2, b). The shrimps apparently shifted their predominant activities to wandering ($34.6 \pm 1.3\%$; bout length = 9.4 ± 0.3 s) and feeding ($24.1 \pm 1.3\%$, bout length ≈ 15 s). Active feeding on particles ($75 \pm 2\%$ of pooled feeding events) was conspicuous; *A. macellarius* only sometimes consumed the hoarded leaves directly and exclusively within the burrow. Both feeding modes were accompanied by either ventilating or grooming bouts, and occurred within 10 and 20 seconds, respectively. Burrowing, i.e., bulldozing bouts in addition to moderate excavating and sweeping events (Table 2, part B), was exhibited for only $15.8 \pm 1.1\%$ – the activity may have ranked second in overall frequency but bout lengths became shorter to half the time observed at initiation (= 7.6 ± 0.3 s). Surveys of burrow walls/ roofs were exhibited less ($7.0 \pm 0.7\%$; bout length = 7.1 ± 0.6 s) and periods of rest, as previously observed, were the least occurring behavior ($1.8 \pm 0.5\%$).

The control of substrate type on behavior was evident in the shrimps' frequencies of burrowing, ventilating, and surveying activities. Burrowing and ventilating were significantly more frequent in sand than in sandy mud (Table 3A, Fig. 5, a) while surveying occurred considerably more often and with extended

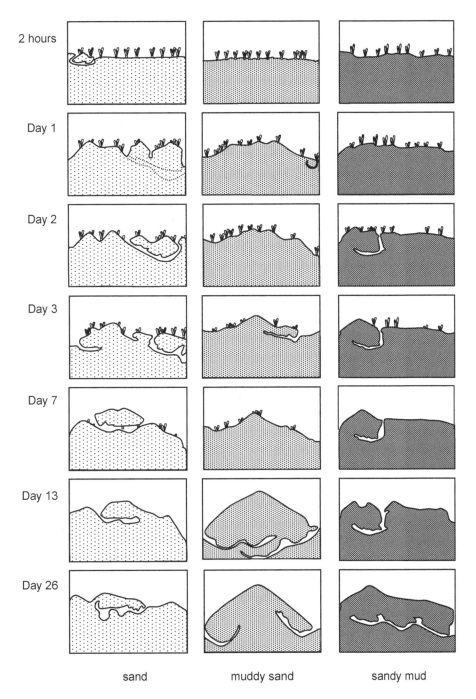

Figure 3. Patterns of burrow excavation by *A. macellarius* depicted after initiation (2 h, Day 0) thru Day 26 in the three substrate types.

periods in muddy sand compared with the other two test substrates (Table 3, A-C, Fig. 5, a-c). In addition to substrate effects, the shrimps exhibited notably more burrowing, grooming, and surveying bouts during the day than at night (Table 3B,

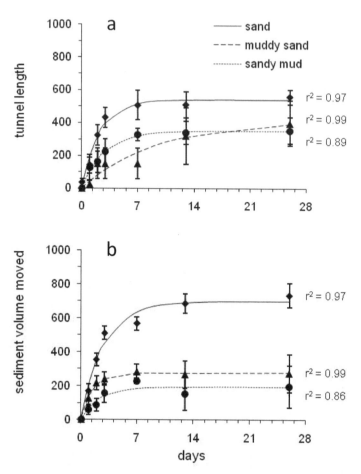

Figure 4. *Alpheus macellarius* turbative activity in the three seagrass substrates (sand, muddy sand, sandy mud) – a, tunnel lengths (mm) and b, volumes of moved sediment (cm³) from initiation up to 26 days in the experimental cuvettes. Means of each parameter in each substrate type were fitted to the function $y = a \times (1 - e^{-b \times t})$.

Fig. 5, a and c). Periods of feeding and wandering, on the other hand, were particularly longer at night than at daytime (Table 3, B and C; Fig. 5, b and c).

Discussion

Burrowing emerged as the principal activity of *Alpheus macellarius* until the first burrow structure was successfully formed, suggesting that the purpose of such intense activity was concealment from the sediment surface. After the shrimps established a first simple hideout, they commenced with tunnel extension and maintenance routines with less intensity, or resolved burrow collapses with reconstruction bouts. This behavior continued up to 1 to 2 weeks, when maximum burrow lengths were achieved and when differences among substrate type became apparent. A similar pattern in the intensity of burrowing behavior with time was

Table 3. Mean squares (MS) and *F*-values from repeated measures ANOVA comparing day- and nighttime behavior of *A. macellarius* (n = 4) in three substrate types. *F*-values with asterisks are significant at $\alpha = 0.05$: * $p<0.05$, ** $p<0.005$, and *** $p<0.001$. Superscripts on the error MS in bout lengths (C, between-subjects effects) denote different *df* values: [a] 43, [b] 38, [c] 67, [d] 55, [e] 38, and [f] 23.

Variable/ Source of variation	Burrow MS	Burrow F	Survey MS	Survey F	Wander MS	Wander F	Feed MS	Feed F	Groom MS	Groom F	Ventilate MS	Ventilate F
A. Frequency (nos. per hour)												
Between-subjects effects:												
Error (within cells; df = 69)	8886.50		2432.80		5502.53		9656.59		931.22		2047.64	
Sediment type (df = 2)	28030.93	3.15*	11298.96	4.64*	311.30	0.06	23311.03	2.41	1047.24	1.12	7025.98	3.98*
Within-subjects effects:												
Error	8015.28		2585.77		5117.76		9423.90		681.92		1732.57	
Time of day (df = 1)	113784.52	14.20***	79891.76	30.90***	1105.78	0.22	13214.34	1.40	3395.16	4.98*	4979.83	2.87
Time of day * Sediment type (df = 2)	3080.85	0.38	2882.10	1.11	1727.91	0.34	5665.34	0.60	103.48	0.15	683.15	0.39
B. Time allocation (%)												
Between-subjects effects:												
Error	261.67		122.24		336.78		427.78		251.64		243.49	
Sediment type	146.75	0.56	713.88	5.84***	746.31	2.22	177.48	0.41	313.21	1.24	662.97	2.72
Within-subjects effects:												
Error	267.67		137.19		341.62		294.20		229.57		172.40	
Time of day	2860.43	10.69**	1886.48	13.80***	3588.53	10.50***	506.89	1.72	576.53	2.51	497.08	2.88
Time of day * Sediment type	96.76	0.36	150.66	1.10	8.91	0.03	113.39	0.39	52.78	0.23	95.42	0.55
C. Bout length (seconds)												
Between-subjects effects:												
Error	154.49[a]		114.50[b]		451.89[c]		464.10[d]		248.62[e]		463.51[f]	
Sediment type	8.74	0.06	376.22	3.29*	291.34	0.64	283.63	0.61	589.12	2.29	187.57	0.40
Within-subjects effects:												
Error	171.15		93.20		392.30		589.71		298.51		186.19	
Time of day	648.51	3.79	11.67	0.13	9475.62	24.20***	3342.74	7.21**	10.46	0.04	439.22	2.36
Time of day * Sediment type	234.99	1.37	97.29	1.04	279.69	0.71	153.89	0.35	148.40	0.50	90.06	0.48

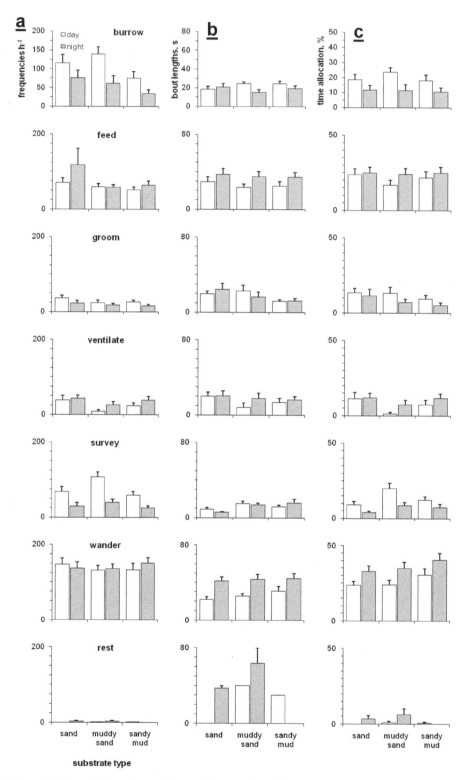

Figure 5. The activity pattern of *A. macellarius* on Day 37: comparisons of measured behavior attributes (a – frequencies, b – bout lengths, c – time allocation) between time of day and among substrate types.

also found in *A. macellarius* observed for five days in outdoor tanks (Palomar et al. 1994, 1995). Burrow diameters appeared similar across the substrates in the constrained cuvettes, although in field conditions, these attributes may have apparent differences due to, e.g., the influence of tides and of exposure (Griffis and Chavez 1988).

The relatively firm and compact matrix of carbonate sand probably contributed to the shrimps' early success in concealment (i.e., within 2 hours) despite the high energy allocated (= large volume of remobilized sediment) in the construction of the burrows. These burrows appeared comparatively stable, thus, presented an opportunity for the shrimps to engage in further burrowing to extend the tunnels so that, eventually, extensive lengths were attained at the end of 4 weeks (54 ± 2 cm). In tank conditions, relatively stable burrows that persisted for five days were observed in setups with 15 and 25% gravel, but that these were likely to collapse in the field because of wave action (Palomar et al. 2005). The interwoven rhizome-root compartment of the dense and mixed seagrass meadows on soft carbonate sand (aboveground and rhizome biomass 346 and 279 g DW m^{-2}, respectively; Vermaat et al. 1995) could provide additional reinforcement to the burrows. These meadows may, thus, be expected to sustain the relatively intense burrowing behavior of *A. macellarius*, which, in due course, would form sufficiently extensive tunnels and significantly more burrow openings aboveground (mean range 3 – 5 m^{-2}, maximum = 15 m^{-2}; Nacorda et al., Chapter 2, this Thesis). In the experiment's terrigenous substrates, the first (simple) burrows were constructed later and tunnels lengths attained at the end of the monitoring period were shorter – *A. macellarius* shifted its efforts to either belowground reconstructions/ new burrow initiation attempts, periods of inactivity once in relatively secure hiding, or to activities exclusively within the burrows (such as the relatively large amount of time spent surveying by shrimps in muddy sand sediments). The frequency of burrow openings in seagrass beds on terrigenous substrates is low (mean range < 1 to 4 m^{-2}; maximum 10 m^{-2} for sand; Nacorda et al., Chapter 2, this Thesis) and may be attributed to substrate instability and the limited stabilizing effect from the roots of the sparse seagrasses (from 172 to 492 shoots m^{-2} in similar environments, Bach et al. 1998). All the leaves of the transplants were removed rapidly and hoarded in the burrows, similar to our observations in the field. The shrimps were also observed to fix buoyant leaves between rubble interstices to prevent dislodgement. Direct feeding on this secured cache was deferred for later, hence, passively subjecting the leaves to 'gardening periods' (Vander Wall 1990) that may improve the leaves' food quality (Andersen and Kristensen 1991). On few occasions the shrimps were observed to directly utilize the fronds, *A. macellarius*, after defecation in the tunnels, contributed to the particulate organic matter pool of the sediment. This organic matter probably increased in digestibility and nutrient content (Thayer et al. 1984), and may be exploited as alternative food. Our observations revealed *A. macellarius* as primarily a deposit feeder, which directly accessed food by scratching/ excavating sediments. The shrimp's deposit feeding mode was clearly associated with burrowing, here represented by the behavioral states 'scratch' and 'excavate' (Table 2, part B), and was observed more frequently after the tunnels had been established and/or extended. The study also revealed that *A. macellarius* is, secondarily, a cryptic suspension feeder, with this feeding mode often occurring in the tunnels – the slow fanning movements of the shrimps' pleopods (state 'vent') suspended detritus particles that were simultaneously caught then consumed. Gut content and stable isotope analyses revealed *A. macellarius* as omnivores, whose diet was composed of largely seagrass and unidentified organic matter, and, to a lesser extent, algae, diatoms, spicules, and crustaceans (Palomar et al. 2004). Feeding behavior in the tanks suggested a deposit-feeding mode, however, and that suspension feeding via the burrow openings was not a vital means to obtain food but rather for burrow ventilation. Other shrimp species demonstrate combined feeding methods to compensate for nutritional deficits and satisfy energy demand (Thayer et al. 1984, Lopez and Levinton 1987, Nickell and Atkinson 1995, Stamhuis et al. 1998). For example, *Callianassa californiensis* and *C. gigas* are able to deposit-feed, utilize *Zostera*, and suspension-feed (Griffis and Chavez 1988), and *Upogebia omissa* filtered suspended particles and directly fed on the sediment, the latter as the main trophic mode (Coelho et al. 2000).

Feeding activity may, over time, limit the burrows' relative stability (e.g., Suchanek 1983). In the field, this probability of unstable burrows would likely be delayed accordingly because of the sediment-stabilizing effect of seagrasses; *A. macellarius* would, however, continually and frequently rework its burrows and then resume with feeding, as was observed in the cuvettes. Further, the shrimps' maintenance routines ensure tunnel stability and preclude tunnel collapse, i.e., during surveying, when walls are inspected and reinforced through tamping, and during sorting, when larger sediment grains are piled or inserted into the walls and tunnel roofs.

Prolonged periods of hiding, based on field observations, were apparent beyond sundown and when strong current episodes were detected at the sediment surface, largely associated with the goby's response to light and water motion conditions at the immediate vicinity of the burrow openings (Karplus 1987). Biological threats – predation and competition – were seen to be anticipated by the goby and communicated to the shrimp prior to both animals' retreat to the burrows for indefinite periods (Karplus 1987, Manthachitra and Sudara 1988). For *A. macellarius* and *C. octafasciatus*, which appear to be obligate partners (Palomar 2002), the latter was observed to remain within a guarding and communication radius of not more than 50 cm from the burrow opening as *A. macellarius* continued with its aboveground activities and maintained its proximity to the substratum. In the experimental setups, *A. macellarius* displayed more active burrowing (and subsequent grooming) and surveying at daytime and continued at night with less intensity; the shrimps also seemed to shift to more feeding and wandering at night. We consider the observed nighttime wandering and continued burrowing without associated feeding as indicators of habituation, especially given that chemical cues from predators and competitors were absent. We presume, however, that the shrimps would limit wandering to occur strictly within the tunnels to minimize the risk of being prey to e.g., eels (Pamintuan and Aliño 1994), and translate these bouts to activities within the goby's 'working' radius aboveground or their belowground tunnels.

This study showed that three known mechanisms of bioturbation (cf. Andersen and Kristensen 1991) were used by *A. macellarius*, and feeding activity was closely associated to each: (1) shallow sediment mixing through bulldozing and sweeping bouts, (2) fecal material production after grazing on seagrass leaves, and (3) ventilation of the burrows through swift or controlled fanning of the pleopods. The control of sediment type on the burrowing behavior and burrows of *A. macellarius* was evident on the first and third mechanisms, which, together with density and the type(s) of seagrass species and the degree by which the symbiotic partners interact in the field, determines the shrimps' spatial distribution within the various environments occupied by seagrasses (Nacorda et al., Chapter 2, this Thesis). The small-scale disturbance which *A. macellarius* populations impose within the top shallow sediment compartment in seagrass meadows is likely to have a major impact on the local environment – from the structure and distribution of biota (de Wilde 1991, Woodin and Marinelli 1991) to system functioning (cf. Meadows 1991).

Acknowledgments. Funding for the study was granted by WOTRO (WB84-413). We thank Criz Ragos and Ronald de Guzman for the unwavering assistance during field- and hatchery work. Nadia Palomar, Boyet Elefante[†], Cris Diolazo, Sheila Albasin, Fe Lomahan, Napo Cayabyab, and Jovy Fomar are acknowledged for their generous help during the cuvette and aquaria set up and behavior filming activities. EJS provided lab space at the Department of Marine Biology (R*u*G) for event recording work.

References

Altmann, J. 1974. Observational study of behavior: sampling methods. Behaviour 49: 227-267.

Andersen, F.Ø. and Kristensen, E. 1991. Effects of burrowing macrofauna on organic matter decomposition in coastal marine sediments. In: Meadows, P.S. and Meadows, A., eds., The environmental impact of burrowing animals and animal burrows. The Proceedings of a Symposium held at the Zoological Society of London, 3-4 May 1990. Symposia of the Zoological Society of London 63: 69-88. Clarendon Press, Oxford.

Atkinson, R.J.A. and Taylor, A.C. 1991. Burrows and burrowing behaviour of fish. In: Meadows, P.S. and Meadows, A., eds., The environmental impact of burrowing animals and animal burrows. The Proceedings of a Symposium held at the Zoological Society of London, 3-4 May 1990. Symposia of the Zoological Society of London 63: 133-182. Clarendon Press, Oxford.

Bach, S.S., Borum, J., Fortes, M.D., and Duarte, C.M. 1998. Species composition and plant performance of mixed seagrass beds along a siltation gradient at Cape Bolinao, The Philippines. Marine Ecology Progress Series 174: 247-256.

Coelho, V.R., Cooper, R.A., and Rodrigues, S.A. 2000. Burrow morphology and behavior of the mud shrimp *Upogebia omissa* (Decapoda: Thalassinidea: Upogebiidae). Marine Ecology Progress Series 200: 229-240.

Frey, R.W. and Howard, J.H. 1975. Endobenthic adaptations of juvenile thalassinidean shrimp. Geological Society of Denmark Bulletin 24: 283-297.

Frouin, P. 2000. Effects of anthropogenic disturbances of tropical soft-bottom benthic communities. Marine Ecology Progress Series 194: 39-53.

Gomez, K.A. and Gomez, A.A. 1984. Statistical procedures for agricultural research, 2nd ed. John Wiley and Sons, Inc., Singapore, 680 p.

Gray, J.H. 1974. Animal-sediment relationships. Oceanography and Marine Biology Annual Reviews 12: 223-261.

Griffis, R.B. and Chavez, F.L. 1988. Effects of sediment type on burrows of *Callianassa californiensis* Dana and *C. gigas* Dana. Journal of Experimental Marine Biology and Ecology 117: 239-253.

Griffis, R.B. and Suchanek, T.H. 1991. A model of burrow architecture and trophic modes in thalassinidean shrimp (Decapoda: Thalassinidea). Marine Ecology Progress Series 79: 171-183.

Jones, C.G., Lawton, J.H., and Schack, M. 1994. Organisms as ecosystem engineers. Oikos 69: 373-386.

Karplus, I. 1987. The association between gobiid fishes and burrowing alpheid shrimps. Oceanography and Marine Biology Annual Reviews 25: 507-562.

Levinton, J. 1995. Bioturbators as ecosystem engineers: control of the sediment fabric, inter-individual interactions, and material fluxes. In: Jones, C.G. and Lawton, C.H., eds., Linking species and ecosystems, pp. 29-36. Chapman and Hall, New York.

Lopez, G.R. and Levinton, J.S. 1987. Ecology of deposit-feeding animals in marine sediments. Quarterly Reviews in Biology 62: 235-260.

Manthachitra, V. and Sudara, S. 1988. Behavioural communication patterns between gobiid fishes and alpheid shrimps. Proceedings of the 6th International Coral Reef Symposium, Guam 2: 769-773.

Martin P. and Bateson, P. 1986. Measuring behaviour: an introductory guide. Cambridge University Press, Cambridge, UK, 200 p.

McManus, J.W., Nañola, C.L. Jr., Reyes, R.B. Jr., and Kesner, K.N. 1992. Resource ecology of the Bolinao coral reef system. ICLARM Studies and Reviews 22, 117 p. International Center for Living Aquatic Resources Management, Manila, Philippines.

Meadows, A. 1991. Burrows and burrowing animals: an overview. In: Meadows, P.S. and Meadows, A., eds., The environmental impact of burrowing animals and animal burrows. The Proceedings of a Symposium held at the Zoological Society of London, 3-4 May 1990. Symposia of the Zoological Society of London 63: 1-13.

Nacorda, H.M.E., Cayabyab, N.M., Fortes, M.D., and Vermaat, J.E. Chapter 2: The distribution of burrowing shrimp disturbance in Philippine seagrass meadows. (this Thesis)

Nacorda, H.M.E., Stamhuis, E.J., and Vermaat, J.E. Chapter 3: Aboveground behavior and significance of *Alpheus macellarius*, Chace, 1988, in a Philippine seagrass meadow. (this Thesis)

Nickell, L.A. and Atkinson, R.J.A. 1995. Functional morphology of burrows and trophic modes of three thalassinidean shrimp species, and a new approach to the classification of thalassinidean burrow morphology. Marine Ecology Progress Series 128: 181-197.

Ott, J.A., Fuchs, B., Fuchs, R., and Malasek, A. 1976. Observations on the biology of *Callianassa stebbingii* Borrodaile and *Upogebia litoralis* Risso and their effect upon the sediment. Senchenbergiana Maritima 8: 61-79.

Palomar, N.E. 2002. The population characteristics and behavior of the burrowing shrimp *Alpheus macellarius* as influenced by sediment conditions and seagrass. M.Sc. Thesis, Marine Science Institute, University of the Philippines, Diliman, Quezon City, 107 p.

Palomar, N.E., Juinio-Meñez, M.A., and Karplus, I. 2004. Feeding habits of the burrowing shrimp *Alpheus macellarius*. Journal of the Marine Biological Association of the United Kingdom 84: 1199-1202.

Palomar, N.E., Juinio-Meñez, M.A., and Karplus, I. 2005. Behavior of the burrowing shrimp *Alpheus macellarius* in varying gravel substrate conditions. Journal of Ethology 23: 173-180.

Pamintuan, I.S. and Aliño, P.M. 1994. The daily food rations of *Gymnothorax* sp. and *Choerogon anchorago* in the reef flat of Bolinao, Pangasinan, Northwestern Philippines. In: Sudara, S., Wilkinson, C.R., and Chou, L.M., eds., 3rd ASEAN-Australia Symposium on Living Coastal Resources, 16-20 May 1994, Chulalongkorn University, Bangkok, Thailand, pp. 269-273. Chulalongkorn University, Bangkok, Thailand.

Potvin, C., Lechowicz, M.J., and Tardiff, S. 1990. The statistical analysis of ecophysiological response curves obtained from experiments involving repeated measures. Ecology 71: 1389-1400.

Stamhuis, E.J., Reede-Dekker, T., van Etten, Y., Wiljes J.J. de, and Videler, J.J. 1996. Behavior and time allocation of the burrowing shrimp *Callianassa subterranea* (Decapoda, Thalassinidea). Journal of Experimental Marine Biology and Ecology 204: 225-239.

Stamhuis, E.J., Schreurs, C.E., and Videler, J.J. 1997. Burrow architecture and turbative activity of the thalassinid shrimp *Callianassa subterranea* from the central North Sea. Marine Ecology Progress Series 151: 155-163.

Stamhuis E.J., Videler, J.J., and Wilde, P.A.W.J. de. 1998. Optimal foraging in the thalassinidean shrimp *Callianassa subterranea*: improving food quality by grain size selection. Journal of Experimental Marine Biology and Ecology 228: 197-208.

Stapel, J.A. and Erftemeijer, P.L.A. 2000. Leaf harvesting by burrowing alpheid shrimps in a *Thalassia hemprichii* meadow in South Sulawesi, Indonesia. Biologia Marina Mediterranea 7: 282-286.

Suchanek, T.H. 1983. Control of seagrass communities and sediment distribution by *Callianassa* (Crustacea, Thalassinidea) bioturbation. Journal of Marine Research 41: 281-298.

Terrados J., Duarte, C.M., Fortes, M.D., Borum, J., Agawin, N.S.R., Bach, S., Thampanya, U., Kamp-Nielsen, L., Kenworthy, W.J., Geertz–Hansen, O., and Vermaat, J. 1998. Changes in community structure and biomass of seagrass communities along gradients of siltation in SE Asia. Estuarine, Coastal and Shelf Science 46: 757-768.

Thayer G.W., Bjorndal, K.A., Ogden, J.C., Williams, S.L., and Zieman, J.C. 1984. Role of larger herbivores in seagrass communities. Estuaries 7: 351-376.

Vander Wall, S.B. 1990. Food hoarding in animals. The University of Chicago Press, Chicago, 445 p.

Vaugelas, J. de. 1985. Sediment reworking by callianassid mud-shrimp in tropical lagoons: a review with perspectives. Proceedings of the 5th International Coral Reef Congress, Tahiti 6: 617-622.

Vermaat, J.E., Agawin, N.S.R., Duarte, C.M., Fortes, M.D., Marbá, N. and Uri, J.S. 1995. Meadow maintenance, growth and productivity of a mixed Philippine seagrass bed. Marine Ecology Progress Series 124: 215-225.

Wilde, P.A.W.J. de. 1991. Interactions in burrowing communities and their effects on the structure of marine benthic ecosystems. In: Meadows, P.S. and Meadows, A., eds., The environmental impact of burrowing animals and animal burrows. The Proceedings of a Symposium held at the Zoological Society of London, 3-4 May 1990. Symposia of the Zoological Society of London 63: 107-117.

Woodin, S.A. and Marinelli, R.: Biogenic habitat modification in marine sediments: the importance of species composition and activity. In: Meadows, P.S. and Meadows, A., eds., The environmental impact of burrowing animals and animal burrows. The Proceedings of a Symposium held at the Zoological Society of London, 3-4 May 1990. Symposia of the Zoological Society of London 63: 231- 250.

Chapter 5

Growth response of the dominant seagrass
Thalassia hemprichii (Ehrenberg) Ascherson
to experimental shrimp disturbance

H.M.E. Nacorda [1, 2], R.N. Rollon [3], M.D. Fortes [1], J.E. Vermaat [2, 4]

[1] Marine Science Institute, University of the Philippines, UPPO Box 1, Diliman, Quezon City, 1101 The Philippines

[2] Department of Environmental Resources, UNESCO-IHE Institute for Water Education, PO Box 3015, 2601 DA Delft, The Netherlands

[3] Institute of Environmental Science and Meteorology, University of the Philippines, Diliman, Quezon City, 1101 The Philippines

[4] present address: Institute for Environmental Studies, Vrije Universiteit, De Boelelaan 1087, 1081 HV Amsterdam, The Netherlands

Abstract

The effects of small-scale disturbance by burrowing shrimps (defoliation, shoot burial) on the growth patterns of the seagrass *Thalassia hemprichii* (Ehrenberg) Ascherson were examined in three manipulative experiments. Disturbance was applied *in situ* on apical shoots (Expt. 1) and on tank-grown seedlings (Expt. 2) and apparent shrimp activity was blocked *in situ* from seagrass plugs in exclosures (Expt. 3). Changes in leaf growth and rhizome elongation/ root growth rates were measured and compared after four short periods (4, 7 or 8, 14, and 25 or 28 days) in Expts. 1 and 2, while in Expt. 3, leaf growth rates of mature *T. hemprichii* shoots and shoot densities of coexisting species within the plugs were monitored for a year. In apical shoots, burial for at least 14 days induced accelerated leaf growth, while clipping and combined treatments had minimal effects on module growth. Clipping did not cause module growth differences with control seedlings. Burial, alone and combined with clipping disturbance, however, significantly and continuously decreased seedling growth. Exclusion from shrimp activity did not influence leaf growth rates of mature shoots. Both *T. hemprichii* growth and shoot densities of other coexisting seagrass species exhibited strong temporal variation as expected.

Our results show the tolerance of vegetative seagrass shoots (both early and late successional stages) to the small-scale disturbance imposed by burrowing shrimps. Seedlings also survived defoliation but were sensitive to burial events.

Keywords: plant response, disturbance, burrowing shrimps, bioturbation, Philippines, SE Asia

Introduction

Burrowing shrimps are important macrofaunal components of tropical seagrass ecosystems (Duarte et al. 1997, Nacorda et al., Chapter 2, this Thesis). In shallow multispecific SE Asian beds, these animals have been observed to rework impressive quantities of sediment (between 0.09 and 1.4 kg m^{-2} d^{-1}) and also harvest living leaf material (between 0.4 to 2.3 g m^{-2} d^{-1} or 12 to 53% of daily biomass production) (Stapel and Erftemeijer 1997, Nacorda et al., Chapter 3, this Thesis). Sediment reworking generates topographic irregularities and gaps in the seagrass canopy, the dimensions and positions of which were found to vary over monthly time scales as the shrimps continually extended their tunnels and relocated their burrow openings and grazing grounds (Nacorda et al., Chapter 3, this Thesis). The harvesting of leaves directly transfers aboveground production to the burrows (Stapel and Erftemeijer 1997), where the material is fractionally consumed by the shrimps (Nacorda et al., Chapter 4, this Thesis) and/ or stored as cache, which undergoes subsequent decomposition (Vander Wall 1990).

The present study focuses on the effects of these two aspects of shrimp behaviour on the common and often dominant seagrass *Thalassia hemprichii* (Ehrenberg) Ascherson. Leaf harvesting has modest to adverse effects on leaf growth (Cebrián and Duarte 1998, Cebrián et al. 1998, Valentine and Heck 1999) and sediment reworking inadvertently results in considerable shoot burial, which may either stimulate or depress seagrass performance (Marbá and Duarte 1994, Marbá et al. 1994, Duarte et al. 1997). We addressed the following questions in a series of three experiments: (a) does leaf clipping and burial increase leaf and rhizome/ root growth in apical shoots and seedlings, and (b) does exclusion from shrimp activity benefit mature shoots and enhance densities of coexisting species? We also assessed whether these disturbance modes interacted in (a), thus, leaf clipping and burial were combined in an *in situ* factorial design.

For Expt. 1, we used undisturbed apical shoots on clonal rhizome runners (lateral meristems; Tomlinson 1974) to address the possible effects of shrimp behaviour on seagrasses established at the edge of bare, shrimp-occupied sand patches. For Expt. 2, we applied the factorial approach on young seedlings grown in an outdoor tank to quantify the sensitivity of newly established plants to shrimp behaviour. The third experiment, conducted *in situ,* was elevated beyond the scale of the shoots and employed experimental exclosures to block all shrimp activity. In this experiment, the effects on growth and shoot densities at the meter-scale in a mixed seagrass bed canopy were monitored over a longer period (up to 1 year). We did the field manipulations in a shallow, clear-water, and lush meadow off Lucero in Bolinao (NW Philippines) where apparent shrimp disturbance was moderate (Nacorda et al., Chapter 2, this Thesis).

Materials and Methods

Experiment 1 – *In situ* leaf clipping and burial of apical shoots of *Thalassia hemprichii*

In August 1999, a total of 360 apical shoots (leaf lengths 33 ± 1 mm; leaf widths 5 ± 0.1 mm) were chosen at random within 40 m² of a mixed seagrass bed on the reef flat off Lucero, Bolinao (16° 24.85' N, 119° 43.48' E). The shoots were tagged, their leaves punched once with a needle below the leaf-sheath conjecture (Zieman 1974), and the lengths and diameters of their respective rhizome apices measured *in situ*. Treatments imposed (= modes of simulated shrimp disturbance) were applied only once on Day 0 on the marked and associated shoots and consisted of leaf clipping (leaves cut right above the needle mark in each shoot), burial at 2 and 5 cm, combined clipping and burial at 2 and 5 cm, and without manipulation (control). Shoots subjected to burial treatments were enclosed with PVC rings (diameter = 32 cm) that were fixed to the substrate prior to the actual burial using ambient mound sediments, i.e., material pumped out and/ or dumped by burrowing shrimps. The setups were left in the field for fixed periods then 10 shoots per treatment were harvested after 4, 8, 14, and 25 days. In the laboratory, blades and rhizomes of the harvested shoots were immediately measured (leaf lengths, widths and growth increments, rhizome lengths and diameters), and the number of marked standing and new leaves were counted.

This experiment was conducted within the window of the rainy season when the zero-datum depth was 0.8 m (tidal range = 1.0 m), the overlying water remained relatively clear (total suspended solids = 2.5 ± 1.0 mg l⁻¹), and temperature and salinity ranged between 29 and 32°C and 30 and 33 psu, respectively (Nacorda, unpubl. data). Surface water velocity during this period averaged 20 cm s⁻¹, which was typical of the rainy season (Rivera et al. 1997).

Experiment 2 – *In situ* leaf clipping and burial of *T. hemprichii* seedlings

In December 1999, bed substrates and mound sediments were collected from four seagrass beds located along the major siltation gradient in Bolinao (Le Jeune 1995, Terrados et al. 1998, Kamp-Nielsen et al. 2002) – Lucero (clear water), Pislatan (16° 22.10' N, 119° 57.72' E), Rufina (16° 21.11' N, 119° 58.03' E), and Dolaoan (most silty; 16° 20.01' N, 119° 57.73' E). These sediments were separately stored; the bed substrates were rid off of plants and decaying matter then set into black seedling pots (8 cm diameter x 10 cm height; n = 2 pots x 4 substrate types x 4 treatments x 5 harvest dates = 160 units prepared in total). Mature fruits of *T. hemprichii* were gathered from Lucero and allowed to dehisce naturally. Testae were removed from the seeds just before planting (n = 5 per pot). Pot units were organized randomly in an outdoor concrete tank (height = 50 cm up to brim) of the Bolinao Marine Laboratory, which continuously received filtered seawater in the daytime. Seeds were left to develop into seedlings for up to a month while algae and epiphytes were carefully brushed off from the units every other day.

Four pot units per substrate type (20 seedlings pooled) were randomly harvested on Day 30 for baseline module measurements – leaf lengths, leaf widths, and root lengths. On the same day, all leaves were marked with a needle hole at 1.3 cm from the base of the seedling then the three modes of simulated disturbance were applied. For the seedlings with the clipping treatments, the leaves were cut at 1.5 cm from the base. Pot units for the burial treatments were wrapped with canvass (Σ height from bottom = 15 cm) then filled up to brim (burial height = 5 cm) with the corresponding mound sediments from the source sites. All the pot units were rearranged randomly within the tank and then maintained as before. Two units per treatment per sediment type (32 units total) were collected on Days 33, 37, 44, and 58, corresponding to 3, 7, 14, and 28 days after the application of experimental disturbance. Post-disturbance sizes of leaves and roots and leaf growth increments were measured after each collection.

Experiment 3 – *In situ* blocking of shrimp disturbance in a *T. hemprichii* meadow

On March 2000, 18 experimental units were set up randomly within 400 m^2 of lush seagrasses in Lucero. These units measured 0.20 to 0.25 m^2 and consisted largely of mature shoots of *T. hemprichii* that were within the immediate vicinity of burrow openings and/ or sand mounds. Thus, all the units have shrimp-disturbed and undisturbed sections, and on these sections. The units were divided into 3 treatments –excluded from shrimp disturbance, procedural controls, and unmanipulated controls. Each excluded unit (= seagrass plug) was sampled using a 50-cm diameter stainless steel corer. This gear was pushed down 20 cm deep into the sediment, stoppered, then slightly lifted as a fine-mesh nylon net cloth (850 μ opening) was slipped underneath it, after which the sediment plug was carefully released back into position. The nets, utilized to block further or potential shrimp activity, extended above the sediment surface (~50 cm) while semi-permanently tied to metal pegs in four corners. Procedural control plugs were similarly set up as the excluded units (but without nets) and were monitored to account for the possible negative effects of core sampling. Control units were simply delimited by metal pegs, into which a 50 x 50 cm quadrat could be directly fitted.

Shoots of all seagrass species in the units were counted. Complete shoots of mature *T. hemprichii* within and beyond 10 cm of an existing shrimp disturbance (n = 5) were also marked. Leaf growth rates were determined from the marked shoots based on *in situ* measurements of leaf lengths and widths. These and shoot densities of all species were monitored in the shrimp-disturbed (or previously disturbed, in the case of the net-excluded units) and undisturbed sections of each unit from March to August 2000 and in April 2001. Nets were periodically cleaned as these accumulated epiphytes and trapped drift algae through time.

Data analyses

Short-term responses of young shoots and seedlings were measured in terms of leaf growth (mm), rhizome elongation/ root growth (mm), and analyzed further as absolute leaf and rhizome growth rates (LAGR and RAGR, respectively, in mm^2 d^{-1}; Rollon 1998), root growth rates, and ratios of module growth (leaf: rhizome growth) and of growth rates (LAGR: RAGR). The overall effect of the disturbance modes and the durations on pairs of correlated dependent variables (Hair et al. 1998), i.e., (a) mean leaf growth and rhizome elongation (r for pooled data = 0.610, $p < 0.001$), (b) LAGR and RAGR ($r = 0.242$, $p < 0.01$), and (c) LAGR and root growth ($r = 0.352$, $p < 0.01$), were evaluated using multivariate analyses of variance (MANOVA). Subsequent MANOVAs were carried out to test for the effect of each independent variable – disturbance mode for each harvest time, and disturbance duration for each disturbance mode – on the pairs of dependent variables. Multivariate differences were based on Pillai's criterion ($\alpha = 0.05$) and where significant probabilities were detected, *post hoc* comparisons (Tukey's HSD) in univariate ANOVA were utilized to identify distinct groups. For the latter analyses, however, α values were adjusted to 0.003 (comparisons of modes) and 0.008 (comparisons of duration) to maintain the experimentwise error rate at 0.05 (Sokal and Rohlf 1995, Hair et al. 1998). The variation in growth rate ratios was evaluated using the Scheirer-Rare-Hare extension of the Kruskal-Wallis test in lieu of the parametric two-way ANOVA (adjusted $\alpha = 0.008$) (Sokal and Rohlf 1995).

For *T. hemprichii* in the field units (Expt. 3), leaf growth rates were similarly calculated as LAGR. Shoot densities and LAGR were compared using repeated measures ANOVA (Sokal and Rohlf 1995) with treatment (exclusion, procedural control, control) as the between-subjects factor and duration of disturbance as the within-subjects factor. The exclosure approach tested the effects of manipulation (exclusion procedure, exclusion of the shrimps) on densities of extant seagrasses.

Results from the univariate tests were used when assumptions of the analyses were met – homogeneous covariance matrices in the between-subjects test and homogeneity of variance in the within-subjects test. When data failed to meet these assumptions, results of multivariate tests (Pillai's trace) were utilized. *Post hoc* comparisons (Tukey's HSD) were carried out on parameters with significant treatment effects ($\alpha = 0.017$).

Results

Experiment 1 – Response of apical shoots to *in situ* leaf clipping and burial

Under background conditions, leaf growth in apical shoots of *Thalassia hemprichii* increased by threefold after 25 days due to new and complete terminal shoots (mean rate 43 ± 5 mm^2 d^{-1}). Rhizome growth increased by fourfold at 5 ± 0.5 mm^2 d^{-1}. Module growth continued under clipping and burial. The response from the leaves, in general, was greater than the response from rhizomes (Fig. 1, a and b). Module growth and absolute growth rates differed significantly in range depending on the mode and duration of the disturbance (Table 1A-1, B-1). The most pronounced differences were due to responses measured 25 days after the application of disturbance (Table 1A-3, 1B-2, 3). Differences in absolute leaf and rhizome growth rate ratios were only significant when disturbance durations were compared (Table 2A, B).

Leaf-clipped shoots maintained lower but steady leaf growth up to 14 days (15 ± 2 mm), and then showed comparable growth with control shoots after 25 days. Leaf growth rates in the disturbed treatments were generally lower than background rates up to Day 14, and then became comparable on Day 25. Leaf growth rates of buried shoots appeared to have accelerated from Days 7 to 25 (Fig. 1, a) while shoots under the combined treatment exhibited a similar range in leaf growth as undisturbed control shoots (Fig. 1, a). High growth rates were also evident in leaves of 2-cm buried shoots (both clipped and unclipped). In clipped shoots with 5-cm burial, leaf and rhizome growth rates slowed down continuously from 67 ± 6 to 40 ± 8 mm^2 d^{-1} and from 8 ± 1 to 3 ± 0.5 mm^2 d^{-1}, respectively (Fig. 1, b). Rhizome extensions were similar for the control and disturbed shoots except those under 5 cm of burial and combined clipping and 2 cm-burial, which had longer lengths of 49 ± 7 and 42 ± 13 mm, respectively, on Day 25.

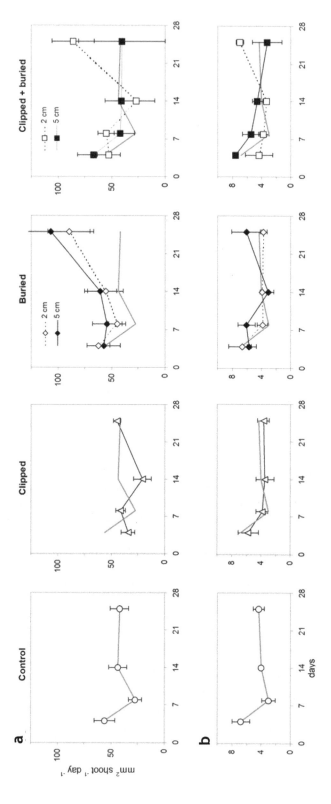

Figure 1. Young *Thalassia hemprichii.* Comparison of leaf growth rates (*a*, as LAGR) and rhizome elongation rates (*b*, as RAGR) (in mean mm^2 d^{-1} ± SEM) between the experimentally disturbed (clipped, buried, clipped + buried) and undisturbed (control, gray lines) apical shoots for each disturbance period (3, 8, 14, and 25 days). (n=10 shoots disturbance mode^{-1} period^{-1})

Table 1A. Multivariate analyses of variance (Expt. 1). The effect of disturbance mode (clipping, burial, and clipping with burial) and disturbance duration (4, 8, 14, 25 days) on leaf growth and rhizome elongation rates in apical shoots of *T. hemprichii* in the field (α = 0.05). Superscripts indicate significant main effects - * $p < 0.05$, ** $p < 0.005$, *** $p < 0.001$ – in leaf ([a]), rhizome ([b]), or both parameters ([c]) from univariate F-tests.

Dependent variable/ MANOVA design/ Source of variation	Pillai's criterion	Approximate F	Hypothesized df	Error df
Mean leaf growth, rhizome elongation (mm)				
1. Two-way factorial				
Duration of disturbance	0.795	29.011 *** [c]	6	264
Disturbance mode	0.173	2.501 * [a]	10	264
Duration x mode	0.326	1.712 ** [b]	30	264
2. One-way factorial: Main effect = disturbance duration				
Clipping	0.713	4.060 *** [c]	6	44
Burial at 2 cm	0.898	5.706 *** [c]	6	42
Burial at 5 cm	1.058	7.108 *** [c]	6	38
Clipping + burial at 2 cm	1.048	7.711 *** [c]	6	42
Clipping + burial at 5 cm	0.658	3.434 * [a]	6	42
Control	0.822	6.515 *** [c]	6	56
3. One-way factorial: Main effect = disturbance mode				
4 days post-disturbance	0.358	1.700	10	78
8 days	0.250	1.027	10	72
14 days	0.332	0.994	10	50
25 days	0.486	2.055 *	10	64

Experiment 2 – Response of seedlings to *in situ* leaf clipping and burial

After 30 days in the outdoor tank, the seedlings of *T. hemprichii* in all the substrates had 4 leaves and 3 roots on average (range = 2 to 6 and 1 to 7, respectively). Leaf material was produced at 30 ± 1 mm^2 d^{-1} while roots lengthened at 6 ± 0.2 mm d^{-1}. Substrate type was not a significant factor in module growth (Pillai's criterion = 0.121, $p > 0.05$).

Within the succeeding 28 days, seedlings displayed significant variation in overall module growth (Fig. 2, a and b), with both disturbance durations and modes particularly affecting leaf growth (Table 3A). In the clipped seedlings, the pattern in leaf growth followed that of the control, although rates appeared slightly lower from Day 37 (= Day 7 post-disturbance) and onwards (Fig. 2, a). Measured rates on Day 58, however, were similar to baseline values (Day 30 or Day 0 post-disturbance; Fig. 2, a). Seedlings with burial and combined treatments decreased leaf growth from Day 33 (20 ± 2 and 27 ± 5 mm^2 d^{-1}, respectively) and by Day 44 (=Day 14 post-disturbance), some recovered seedlings were smothered and growth rates in both treatments became significantly depressed (7 ± 3 and 4 ± 1 mm^2 d^{-1},

Table 1B. Multivariate analyses of variance (Expt. 1). The effect of disturbance mode (clipping, burial, and clipping with burial) and disturbance duration (4, 8, 14, 25 days) on absolute leaf growth (LAGR) and rhizome elongation rates (RAGR) in young *T. hemprichii* in the field ($\alpha = 0.05$). Superscripts indicate significant main effects ([*]$p < 0.05$, [**]$p < 0.005$, [***]$p < 0.001$) in leaf ([a]) or both leaf and rhizome parameters ([b]) from univariate F-tests.

Dependent variable/ MANOVA design/ Source of variation	Pillai's criterion	Approximate F	Hypothesized df	Error df
LAGR, RAGR				
1. Two-way factorial				
Duration of disturbance	0.219	5.409 *** [b]	6	264
Disturbance mode	0.159	2.278 * [a]	10	264
Duration x mode	0.261	1.321	30	264
2. One-way factorial: Main effect = disturbance duration				
Clipping	0.470	2.251	6	44
Burial at 2 cm	0.431	1.925	6	42
Burial at 5 cm	0.426	1.714	6	38
Clipping + burial at 2 cm	0.177	2.195	6	42
Clipping + burial at 5 cm	0.464	2.114	6	42
Control	0.310	1.714	6	56
3. One-way factorial: Main effect = disturbance mode				
4 days post-disturbance	0.204	0.887	10	78
8 days	0.279	1.168	10	72
14 days	0.332	0.996	10	50
25 days	0.639	3.006 ** [a]	10	64

respectively). Such low rates continued until the end of the observation period (Fig. 2, a).

The response of the roots was not as considerable as the response of the leaves (Fig. 2, b) but the significant effects of disturbance mode and duration were still pronounced (Table 3B, C). Growth rates decreased continuously in the buried and in the combined clipped and buried seedlings, i.e., from 6 ± 0.3 to 2 ± 0.3 mm d^{-1} (Fig. 2, b). In contrast (except for the drop on Day 37 or Day 7 post-disturbance) root growth of clipped seedlings was similar to that of the control (pooled mean = 5.2 ± 0.3 mm d^{-1}, Fig. 2, b).

Experiment 3 – Response of excluded seagrasses to blocked shrimp activity

Leaf growth rates of *T. hemprichii* in the excluded and procedural control units were similar with background rates ($p > 0.05$; Table 4, A; Fig. 3) regardless of the distance of the shoots from shrimp disturbance, but appeared strongly seasonal ($p < 0.001$, Table 4). Highest growth rates were found in March and April (169 ± 70 mm^2 shoot^{-1} d^{-1} maximum) and lowest in July (41 ± 6 mm^2 shoot^{-1} d^{-1} minimum) (Fig. 3).

Table 2. Summary of non-parametric two-way analyses of variance (Scheirer-Rare-Hare extension of the Kruskal-Wallis tests) on ranked ratios of mean leaf growth and rhizome elongation (A) and LAGR and RAGR (B). Note: [a] computed as $\Sigma SS / \Sigma df$ due to the presence of ties in the ranks, or [b] otherwise, as $[n(n+1)] / 12$; [*] significant at $\alpha = 0.008$

Dependent variables/ Source of variation	Sum of Squares	df	Computed H	Critical χ^2 ($\alpha = 0.05$)
A. Mean leaf growth: *r*hizome elongation				
Duration of disturbance	8058.304	3	4.039	7.815
Disturbance mode	5753.303	5	2.884	11.070
Duration x mode	29031.862	15	14.553	24.996
Error (replicates)	266375.787	132		
Total	309219.256	155		
Mean Square [a]	1994.963			
B. LAGR: RAGR				
Duration of disturbance	16881.522	3	8.064[*]	7.815
Disturbance mode	11050.359	5	5.278	11.070
Duration x mode	18546.453	15	8.859	24.996
Error (replicates)	272754.185	132		
Total	319232.519	155		
Mean Square [b]	2093.500			

Four other species occurred with *T. hemprichii* in the experimental units – *Halophila ovalis* (R. Br.) Hook, *Cymodocea rotundata* Ehrenb. and Hempr. *ex* Aschers., *Halodule uninervis* (Forssk.) Aschers., and *Enhalus acoroides* (L.f.) Royle. Except in *C. rotundata*, shoot densities varied significantly throughout the observation period ($p < 0.05$ for month; Table 4, B; Fig. 4). The negative effects of exclusion and core sampling on shoot densities were also apparent in some months ($p < 0.05$ for treatment; Table 4, B), since background densities appeared higher in most cases ($p < 0.017$, Tukey's HSD).

Discussion

The modes of applied experimental disturbance elicited both positive and negative responses in shoots of *Thalassia hemprichii*. Leaf harvesting was not damaging while burial appeared to be beneficial to established shoots. The employed drastic clipping regime (>36%, cf. Cebrián and Duarte 1998), which was applied to mimic the leaf harvesting behaviour of burrowing alpheid shrimps, did not lead to shoot mortality or prolonged reduction in leaf and root growth of both apical shoots and seedlings. Single partial defoliation events were found to have little impact on leaf growth in this species (Cebrián and Duarte 1998) and its congener *Thalassia testudinum* (Tomasko and Dawes 1989a). This has been attributed to translocation processes of carbohydrates from the rhizomes (Dawes and Lawrence 1979; Tomasko and Dawes 1989b) and of leaf and root nitrogen (Valentine et al. 2004) to the affected shoots, where the basal meristem can simply continue producing new leaves.

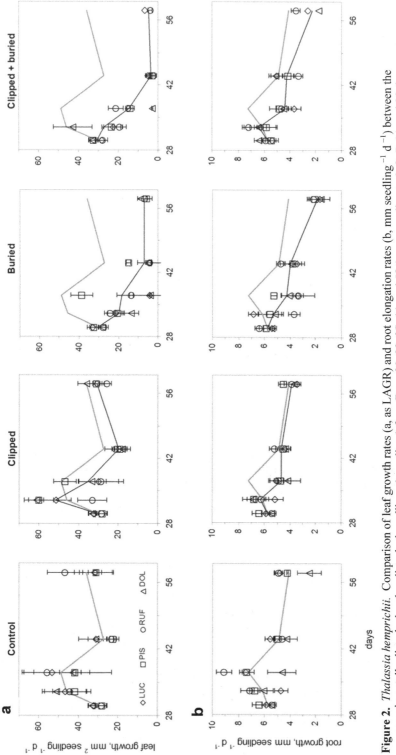

Figure 2. *Thalassia hemprichii.* Comparison of leaf growth rates (a, as LAGR) and root elongation rates (b, mm seedling^{-1} d^{-1}) between the experimentally disturbed and undisturbed seedlings (gray lines) from Days 30, 33, 37, 44, and 58 (corresponding to 0, 3, 7, 14, and 28 days post-disturbance). (n = 10 seedlings disturbance mode^{-1} period^{-1}; different symbols denote sediment sources along the siltation gradient in Bolinao: LUC-Lucero (clear-water, carbonaceous sand), PIS-Pislatan, RUF-Rufina (silty, muddy sand), DOL-Dolaoan (most silty, sandy mud)

Table 3. Multivariate analyses of variance (Expt. 2). The effect of disturbance and disturbance duration (4, 7, 14, 28 days) on the correlated parameters (LAGR, root lengths) measured in *T. hemprichii* seedlings ($\alpha = 0.05$). Superscripts indicate significant main effects ([**]$p < 0.005$, [***]$p < 0.001$) in leaf ([a]), root ([b]), or both measures ([c]) from univariate *F*-tests.

MANOVA design/ Source of variation	Pillai's criterion	Approximate *F*	Hypothesized *df*	Error *df*
A. Two-way factorial				
Duration of disturbance	0.236	13.420 *** [a]	8	802
Disturbance mode	0.095	6.671 *** [a]	6	802
Duration x mode	0.143	3.448 *** [c]	18	802
B. One-way factorial: Main effect = disturbance duration				
Clipping	0.495	9.430 *** [c]	6	172
Burial (5 cm)	0.276	3.685 ** [a]	6	138
Clipping + burial	0.638	10.463 *** [c]	6	134
Control	0.277	5.081 *** [c]	6	190
C. One-way factorial: Main effect = disturbance mode				
3 days post-disturbance	0.419	8.044 *** [a]	6	187
7 days	0.520	8.669 *** [c]	6	148
14 days	0.484	9.905 *** [a]	6	186
28 days	0.577	7.981 *** [c]	6	118

Small-scale and short-term burial resulted in a different response in established shoots and seedlings by stimulating leaf growth by up to 2.6% in the former (mean range from 1.1 to 1.5%) but reducing growth by as much as 15% in the latter (mean range from 1.7 to 7.7%). The enhanced growth response agrees with previous observations of induced compensatory growth in apical shoots, i.e., that increased leaf production and longer new internodes were found (Gallegos et al. 1993, Marbá and Duarte 1994, Duarte et al. 1997) at intermediate burial levels for <2 months (cf. Vermaat et al. 1997). Rapid growth of seagrass leaves can be sustained by the translocation of reserve carbohydrates and nutrients (Terrados et al. 1997, Uy et al. 2001a), which are present in substantial quantities in the robust rhizomes of *T. hemprichii* (about 40% of the rhizome DW; Vermaat et al., unpubl. data), by reductions in the transport of leaf photosynthates to the rhizomes (Uy et al. 2001b) and possible reallocations of energy and resources from carbon assimilation to light harvesting, as was demonstrated for *T. testudinum* (Major and Dunton 2002). In contrast, the one month-old seedlings failed to recover, even at intermediate burial depths for short-term periods (Fig. 2), as their nutritional reserves were probably insufficient to meet the demand for increased leaf production that was necessary to counteract the negative effects of burial. This growth reduction was also apparent when burial was combined with one-time defoliation. Growth of established shoots under combined disturbance neither improved nor declined (Fig. 1, Table 1B-2) although the tendency towards recovery was apparent after 25 days in shoots buried in 2 cm of sediment. Carbohydrate reallocation to clipped shoots may have lasted about 25 days with 2 cm burial, but in clipped shoots with twice the burial depth, we presume that translocations may have proceeded considerably longer.

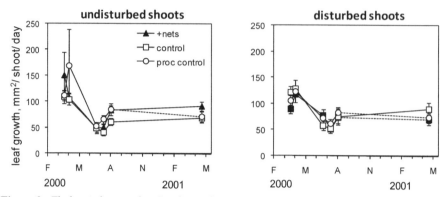

Figure 3. *Thalassia hemprichii.* Leaf growth rates (mean ± SEM) of mature shoots in the undisturbed and disturbed sections of the experimental units – excluded (+ nets), control, procedural (proc) control – from March to August 2000 and in April 2001.

The ranges and the temporal pattern observed for leaf growth (Fig. 3, Table 4-A) and shoot densities of *T. hemprichii* vegetative shoots (Fig. 4, Table 4-B) were consistent with previous observations (Rollon 1998, Agawin et al. 2001). This emphasizes the significance of environmental forcing (cloudiness, rainfall, water turbulence, and total daily PAR) over species-specific traits. The increase in seagrass densities within the control plots was not observed in the enclosures. This indicates successful recruitment of vegetative shoots outside the exclosures and suggests that our barrier against shrimp activity may also have prevented clonal recruitment of most long-lived species. Shoots of *Halophila ovalis*, on the other hand, increased in numbers despite the barrier, and successfully colonized the disturbed section of the exclosures (Fig. 4), as expected for pioneer species (Brouns 1987, Clarke and Kirkman 1989, Duarte et al. 1997).

Small gaps in the canopy continuously created by burrowing shrimps, thus, become sites for fast clonal growth (Rollon et al. 1998, Rasheed 1999, Olesen et al. 2004), and, probably, for the recruitment of seedlings (Dumbauld and Wyllie-Echeverria 2003, Olesen et al. 2004). Although continued shrimp disturbance in these patches may limit the survival of these sexual recruits (Lacap et al. 2002, Balestri and Cinelli 2003, Rollon et al. 2003, Olesen et al. 2004), the gaps maintain small-scale heterogeneity in the landscape of shallow mixed beds, which allows for a suite of different species to coexist (Duarte et al. 1997) in a situation of continuous, small-scale recolonization (Rollon et al. 1998, Rasheed 1999, Olesen et al. 2004). The high sensitivity of *Thalassia* seedlings to both clipping and burial may be an important reason for the high mortality rates of these young recruits. Although massive seedling recruitment can sometimes be observed sometimes (Olesen et al. 2004), seedling densities are often very low. Activities of the often abundant alpheid shrimps are thus probable causes of seedling loss. This, and the availability of new bare habitat, may well have selected for a well-developed clonal expansion capacity in the suite of seagrass species present.

Acknowledgments. WOTRO, through Project WB84-413, provided funding for the study and the Marine Environment Resources Foundation, Inc. (MERF) granted a writing scholarship contract to HEN. We are grateful to Cris Diolazo, who helped in

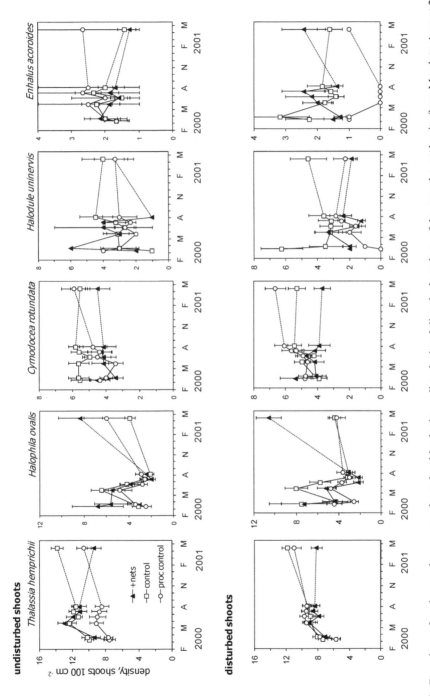

Figure 4. Experiment 3 – changes in seagrass shoot densities in the undisturbed and disturbed sections of the experimental units from March to August 2000 and in April 2001.

Table 4. Repeated measures ANOVA (Expt. 3). The effects of exclusion and sampling procedure ([a]) and distance from shrimp disturbance ([b]) on leaf growth rates of *T. hemprichii* (A) and on seagrass densities (B) in the experimental units. The main effect of month (within-subjects factor) and the interaction terms were evaluated from results of the multivariate tests ([c] as Pillai's trace). Significant F-values are ($\alpha = 0.05$) are indicated by asterisks : * $p < 0.05$, ** $p < 0.005$, *** $p < 0.001$.

Variables	Error (w/in replicates)	Between-subjects effects			Within-subjects effects[c]			
		Treatment[a] (T, df=2)	Shoot location[b] (SL, df=1)	T x SL (df=2)	Month, M	M x T	M x SL	M x T x SL
A. LAGR, mm²/ shoot/ d								
MS	0.76 (df=88)	0.82	0.23	0.01	0.46 (df=5)	0.11 (df=10)	0.05 (df=5)	0.06 (df=10)
F		1.10	0.30	0.10	14.45 ***	0.43	0.82	0.56
B. Seagrass densities, nos./ 100 cm²								
Thalassia hemprichii								
MS	90.85 (df=193)	597.41	202.96	470.87	0.31 (df=6)	0.14 (df=12)	0.16 (df=6)	0.08 (df=12)
F		6.57 ***	2.23	5.18 **	13.71 ***	2.45 **	5.78 ***	1.30
Halophila ovalis								
MS	25.99	82.18	28.00	52.51	0.37	0.30	0.04	0.09
F		3.16 *	1.08	2.02	18.74 ***	5.67 ***	1.49	1.48
Cymodocea rotundata								
MS	50.61	431.14	100.11	173.42	0.56	0.15	0.11	0.06
F		8.52 ***	1.98	3.43 *	1.81	2.61 ***	3.99 ***	0.95
Halodule uninervis								
MS	5.87	35.06	2.05	1.33	0.07	0.63	0.02	0.14
F		5.97 ***	0.35	0.23	2.48 *	1.03	0.72	2.35 **
Enhalus acoroides								
MS	3.27	2.19			0.07	0.19	0.02	0.11
F		0.67			2.24 *	0.30	0.82	1.79 *

the fabrication of the field corer, to staff, researchers, and volunteers at BML who assisted during experimental manipulations, field/ hatchery retrievals, and subsequent measurements for the duration of the study – Cris Ragos, Ronald de Guzman, Oytón Rubio Jr., Chà de Castro, Mímoy Silvano, Ronald Gijlstra, José Vos, Kat Villamor, Rex Montebon, Rómmi Dizon, Makóy Ponce, Frank Wiegman, and Sheila Albasin. Dr. Dosette Pante explained patiently on the difficult details of multivariate statistics.

References

Agawin, N.S.R., Duarte, C.M., Fortes, M.D., Uri, J.S., and Vermaat, J.E. 2001. Temporal changes in the abundance, leaf growth and photosynthesis of three co-occurring Philippine seagrasses. Journal of Experimental Marine Biology and Ecology 260: 217-239.

Balestri, E., and Cinelli, F. 2003. Sexual reproductive success in *Posidonia oceanica*. Aquatic Botany 75: 21-32.

Brouns, J.J.W.M. 1987. Growth patterns in some Indo-West Pacific seagrasses. Aquatic Botany 28: 39-61.

Cebrián, J., and Duarte, C.M. 1998. Patterns in leaf herbivory on seagrasses. Aquatic Botany 60: 67-82.

Cebrián, J., Duarte, C.M., Agawin, N.S.R., and Merino, M. 1998. Leaf growth response to simulated herbivory: a comparison among seagrass species. Journal of Experimental Marine Biology and Ecology 220: 67-81.

Clarke, S.M. and Kirkman, H. 1989. Seagrass dynamics. In: Larkum, A.W.D., McComb, A.J., and Shepherd, S.A., eds., Biology of seagrasses, pp, 304-345. Elsevier, Amsterdam.

Dawes, C.J., and Lawrence, J.M. 1979. Effects of blade removal on the proximate composition of the rhizome of the seagrass *Thalassia testudinum* Banks *ex* König. Aquatic Botany 7: 255-266.

Duarte, C.M., Terrados, J., Agawin, N.S.R., Fortes, M.D., Bach, S., and Kenworthy, W.J. 1997. Response of a mixed Philippine seagrass meadow to experimental burial. Marine Ecology Progress Series 147: 285-294.

Dumbauld, B.R., and Wyllie-Echeverria, S. 2003. The influence of burrowing thalassinid shrimps on the distribution of intertidal seagrasses in Willapa Bay, Washington, USA. Aquatic Botany 77: 27-42.

Gallegos, M., Merino, M., Marba, N., and Duarte, C.M. 1993. Biomass and dynamics of *Thalassia testudinum* in the Mexican Caribbean: elucidating rhizome growth. Marine Ecology Progress Series 95: 185-192.

Hair, Jr., J.F., Anderson R.E., Tatham, R.L., and Black, W.C. 1998. Multivariate data analysis, 5th ed. Prentice Hall International, Inc., New Jersey, 730 p.

Kamp-Nielsen, L., Vermaat, J.E., Wesseling, I., Borum, J., and Geertz-Hansen, O. 2002. Sediment properties along gradients of siltation in South-east Asia. Estuarine, Coastal and Shelf Science 54: 127-137.

Lacap, C.D.A., Vermaat, J.E., Rollon, R.N., and Nacorda, H.M. 2002. Propagule dispersal of the S.E. Asian seagrasses *Enhalus acoroides* and *Thalassia hemprichii*. Marine Ecology Progress Series 235: 75-80.

Le Jeune, E.L.: 1995. Causes of siltation in the Santiago Island reef system. Wagenigen Agricultural University.

Major, K.M., and Dunton, K.H. 2002. Variations in light-harvesting characteristics of the seagrass, *Thalassia testudinum*: evidence for photoacclimation. Journal of Experimental Marine Biology and Ecology 275: 173-189.

Marbá, N., and Duarte, C.M. 1994. Growth response of the seagrass *Cymodocea nodosa* to experimental burial and erosion. Marine Ecology Progress Series 107: 307-311.

Marbá, N., Gallegos, M.E., Merino, M., and Duarte, C.M. 1994. Vertical growth of *Thalassia testudinum*: seasonal and interannual variability. Aquatic Botany 47: 1-11.

Nacorda, H.M.E., Cayabyab, N.M., Fortes, M.D., and Vermaat, J.E. Chapter 2: The distribution of burrowing shrimp disturbance in Philippine seagrass meadows. (this Thesis)

Nacorda, H.M.E., Stamhuis, E.J., and Vermaat, J.E. Chapter 3: Aboveground behavior and significance of *Alpheus macellarius*, Chace, 1988, in a Philippine seagrass meadow. (this Thesis)

Nacorda, H.M.E., Stamhuis, E.J., and Vermaat, J.E. Chapter 4: Burrows and behavior of the snapping shrimp *Alpheus macellarius* Chace, 1988, in different seagrass substrates. (this Thesis)

Olesen, B., Marbá, N., Duarte, C.M., Savella, R.S., and Fortes, M.D. 2004. Recolonization dynamics in a mixed seagrass meadow: the role of clonal versus sexual processes. Estuaries 27: 770-780.

Rasheed, M.A. 1999. Recovery of experimentally created gaps within a tropical *Zostera capricornii* (Aschers.) seagrass meadow, Queensland, Australia. Journal of Experimental Marine Biology and Ecology 235: 183-200.

Rivera, P.C. 1997. Hydrodynamics, sediment transport and light extinction off Cape Bolinao, Philippines. PhD thesis, Wageningen Agricultural University and IHE-Delft. A.A. Balkema, Rotterdam, 244 p.

Rollon, R.N. 1998. Spatio-temporal variation in shoot size and leaf growth of the two dominant Philippine seagrasses *Enhalus acoroides* (L.f.) Royle and *Thalassia hemprichii* (Ehrenb.) Aschers. In: Rollon, R.N., Spatial variation and seasonality in growth and reproduction of *Enhalus acoroides* (L.f.) Royle populations in the coastal waters off Cape Bolinao, NW Philippines, pp. 37-52. PhD Thesis, Wageningen Agricultural University and IHE-Delft. A.A. Balkema, Rotterdam.

Rollon, R.N., de Ruyter van Steveninck, E.D., Van Vierssen, W., and Fortes, M.D. 1998. Contrasting recolonization strategies in multi-species seagrass meadows. Marine Pollution Bulletin 137: 450-459.

Rollon, R.N., Vermaat, J.E., and Nacorda, H.M.E. 2003. Sexual reproduction in SE Asian seagrasses: the absence of a seed bank in *Thalassia hemprichii*. Aquatic Botany 75: 181-185.

Sokal, R.R., and Rohlf, F.J. 1995. Biometry, the principles and practice of statistics in biological research, 3rd ed. W.H. Freeman and Co., New York, 887 p.

Stapel, J., and Erftemeijer, P.L.A. 1997. Leaf harvesting and sediment reworking by burrowing alpheid shrimps in a *Thalassia hemprichii* meadow in South Sulawesi, Indonesia. In: Stapel, J., Nutrient dynamics in Indonesian seagrass beds: factors determining conservation and loss of nitrogen and phosphorus, pp. 33-41. PhD thesis, Katolieke Universitaet Nijmegen. WOTRO/ NWO, The Hague.

Terrados, J., Duarte, C.M., Fortes, M.D., Borum, J., Agawin, N.S.R., Bach, S., Thampanya, U., Kamp-Nielsen, L., Kenworthy, W.J., Geertz-Hansen, O., and Vermaat, J.E. 1998. Changes in community structure and biomass of seagrass communities along gradients of siltation in SE Asia. Estuarine, Coastal and Shelf Science 46: 757-768.

Terrados, J., Duarte, C.M., and Kenworthy, W.J. 1997. Is the apical growth of *Cymodocea nodosa* dependent on clonal integration? Marine Ecology Progress Series 158: 103-110.

Tomasko, D.A., and Dawes, C.J. 1989a. Effects of partial defoliation on remaining intact leaves in the seagrass *Thalassia testudinum* Banks *ex* König. Botanica Marina 32: 235-240.

Tomasko, D.A., and Dawes, C.J. 1989b. Evidence for physiological integration between shaded and unshaded short shoots of *Thalassia testudinum*. Marine Ecology Progress Series 54: 299-305.

Tomlinson, P.B. 1974. Vegetative morphology and meristem dependence – the foundation of productivity in seagrasses. Aquaculture 4: 107-130.

Uy, W.H., Hemminga, M.A., and Vermaat, J.E. 2001a. The effects of shading on clonal integration in the seagrasses *Thalassia hemprichii* and *Halodule uninervis*. In: Uy, W.H., Functioning of Philippine seagrass species under deteriorating light conditions, pp. 73-92. PhD thesis, Wageningen Agricultural University and IHE-Delft. Swets and Zeitlinger B.V., Lisse.

Uy, W.H., Hemminga, M.A., and Vermaat, J.E. 2001b. The effects of long-term light reduction on carbon allocation in selected seagrass species using the stable isotope [13]C. In: Uy, W.H., Functioning of Philippine seagrass species under deteriorating light conditions, pp. 35-47. PhD thesis, Wageningen Agricultural University and IHE-Delft. Swets and Zeitlinger B.V., Lisse.

Valentine, J.F., and Heck Jr., K.L. 1999. Seagrass herbivory: evidence for the continued grazing of marine grasses. Marine Ecology Progress Series 176: 291-302.

Valentine, J.F., Fennel Blythe, E., Madhavan, S., and Sherman, T.D. 2004. Effects of simulated herbivory on nitrogen enzyme levels, assimilation and allocation in *Thalassia testudinum*. Aquatic Botany 79: 235-255.

Vander Wall, S.B. 1990. Food hoarding in animals. The University of Chicago Press, Chicago, 445 p.

Vermaat, J.E., Agawin, N.S.R., Fortes, M.D., Uri, J.S., Duarte, C.M., Marbá, N., Enriquez, S., and van Vierssen, W. 1997. The capacity of seagrasses to survive increased turbidity and siltation: the significance of growth form and light use. Ambio 26: 499-504.

Zieman, J.C. 1974. Methods for the study of the growth and production of turtle grass, *Thalassia testudinum* konig. Aquaculture 4: 139-143.

Chapter 6

General discussion and conclusions

Because the footprint of man is now perceptible in all aspects of the state of the natural world, such that we accept that we are now in the Era of the "Anthropocene" (Crutzen and Stoermer 2000), there is much emphasis on quantifying the impact of global change on biological distributions, biodiversity, and ecosystem functions (Schulze and Mooney 1994, Chapin et al. 1997, Snelgrove 1999, Duarte 2000, Hughes 2000, Smith et al. 2000, Harley et al. 2006, Occhipinti-Ambrogi 2007). The discussion of Hughes (2000) however, recommended that these types of studies need to be grounded on sound monitoring of how existing natural perturbations already influence the behavior and life cycles of flora and fauna of interest. The series of papers compiled by Pickett and White (1985) on *The Ecology of Natural Disturbance and Patch Dynamics* and by Hobbie et al. (1994) are examples of good background as basis to study the effects of global change over and above natural disturbances. Changes in global climate are anticipated to greatly impact the world's seagrasses (Short and Neckles 1999). The responses are predicted to be linked with hydrodynamics, which affects every aspect of the existence of seagrasses, the benthos in general (Koch et al. 2006), and the processes at various scales that affect clonal growth, and eventually, seagrass colonization and recovery from disturbance (Hemminga and Duarte 2000, Duarte et al. 2006). One focus of global change studies is watershed modification and sediment delivery to the coast and its consequences to coastal habitat distribution and health (e.g., Terrados et al. 1998, Kamp-Nielsen et al. 2002, Orth et al. 2006). Studies on the natural reworking of sediments in seagrass meadows have also developed from providing evidence of biotic interactions to the perspective of ecosystem engineering (e.g., Berkenbusch et al. 2000, Siebert and Branch 2006, Berkenbusch et al. 2007, Pillay et al. 2007).

This thesis covered one type of natural disturbance, small-scale, but ubiquitous in dense SE Asian seagrass beds. The work only barely scraped the surface of interactions that can be anticipated when an integrated approach to studying burrowing animals is coupled with analyses of associations with other fauna (mutualisms), the properties of the seagrass canopy and sediments, and the resultant seagrass and endofaunal distributions. The major issues addressed in the thesis are highlighted and integrated in this chapter, particularly describing the bioturbation of burrowing shrimps, the foraging of *Alpheus macellarius* and its mutualism with the species of the goby genus *Cryptocentrus*, the response of the dominant *Thalassia hemprichii* to their small-scale disturbance, and the implications of all these to large-scale seagrass dynamics. Finally, future directions are suggested.

Bioturbation by burrowing shrimps in seagrass meadows

Burrowing is a form of bioturbation, which mixes sediments and porewater and allows for frequent release of nutrients from belowground to the sediment surface, making them more readily available to benthic primary producers in general. This study showed that the burrowing shrimps *A. macellarius* (Caridea) and species of Thalassinidea are major agents of bioturbation in shallow Philippine seagrass meadows, mainly gauged from the size, abundance, and distribution of sediment gaps (Chapter 2). These animals are mainly engaged with burrowing or mining, with reworking rates of between 8 and 1.4 kg m^{-2} d^{-1} for *A. macellarius* (Chapter 3) and up to 12 kg m^{-2} d^{-1} for the thalassinidean shrimps. A potential follow-on study may assess the impact of these dominant burrowing shrimps to the environment based on other indicators. Meadows (1991b) has proposed indices of complexity (*C*), tortuosity (*T*), and activity (*A*), all morphological descriptions that may be tested and all independent of scale, so that impact and modification in terrestrial and aquatic habitats may be compared regardless of the sizes of animals involved. Moreover, the counterparts and functional roles in Philippine seagrass beds of temperate macrofaunal organisms that cause significant bioturbation, e.g., lugworms (Philippart 1994), crabs, and rays (Woods and Schiel 1997, Townsend and Fonseca 1998), remain to be assessed and evaluated at appropriate scales.

Burrowers both affect and are affected by mechanical properties of sediments (Dorgan et al. 2006). We showed in Chapter 4 that soft carbonate sand sediments present greater support for burrowing behavior and burrow complexity and, thus, concealment of *A. macellarius* than terrigenous seagrass substrates, where burrowing appears limited and substituted by concealment strategies, periods of inactivity, or within-burrow activities. The rhizome-root compartment of the dense and mixed seagrass bed often in soft carbonate sand also provides additional reinforcement to the burrows, and the beds, therefore, may be expected to sustain the intense burrowing behavior of *A. macellarius*. Hence, significantly more burrow openings were found (Chapter 2). The reduced diversity and densities of seagrasses in silty beds appear to offer less support to the burrowing behavior of *A. macellarius*, and, in parallel, the frequency of burrow openings was observed to be low (Chapter 2).

On the larger scale, hydrodynamics was important in constraining the spatial distribution of shrimp disturbance in the meadows, i.e., sediment gaps were more pronounced in sheltered than in open beds (Chapter 2). Hydrodynamics also influenced the periodicity of the aboveground activity of alpheid shrimps, which were found to be more prominent in the dry than during the wet months (Chapter 3). It would be interesting in future to quantify the flow properties and hydrodynamic forces operating at the sediment-water interface and within- and above the canopy that allow or restrict the aboveground activities of *A. macellarius* and, likewise, determine the distribution of faunal disturbance. Koch et al. (2006) discussed the increase of turbulence and mixing within canopies of less shoots or where seagrass patchiness is increased. In the studied seagrass habitats associated with the presence of rivers, water-column K_d was also high and the presence of shrimp gaps was observed to be limited (Chapter 2). High K_d may account for observed patchiness of seagrasses in the turbid areas visited and is also linked to sediments that are easily resuspended (e.g., Rollon 1998), and burrowing alpheid shrimps responded to these by spending more time hidden in their burrows. The (*in situ*) aboveground activity of *A. macellarius* in these seagrass areas still needs to be established and compared with *A. macellarius* in clear-water beds.

Trophic strategies of *A. macellarius*

Burrowing and the other mechanisms of bioturbation by *A. macellarius* are closely associated with its feeding (Chapter 4). Mound-forming thalassinidean shrimps are deposit feeders (Griffis and Suchanek 1991, Nickell and Atkinson 1995) and *A. macellarius* demonstrated a similar mode as its main feeding strategy (Chapter 4, this Thesis; Palomar et al. 2004) combined with grazing and suspension feeding (Chapter 4). *Alpheus macellarius* harvests seagrass leaves, hoards the harvest belowground (Chapter 3) and consumes these in its burrows (Chapter 4). Leaf harvesting appears to limit the shrimps' distribution to seagrass-vegetated substrates (Chapter 2). Food hoarding, as an adaptive strategy, gains advantage for *A. macellarius* in the control of food availability in space and during environmental contingencies, e.g., when water movement is high (storms) or light is limiting (turbid) and emergence at the surface is perceived as risky. It also implies that the sediment surface of seagrass beds presents a risk for their foraging ('security hypothesis', Vander Wall 1990). Thus, because of its leaf hoarding behavior and ability to mix feeding strategies (grazing, deposit- and suspension-feeding; Chapter 4), *A. macellarius* demonstrates its fitness in a meadow that does not appear to be resource-limited. The potential food limitation by leaf availability can only be assessed, however, on a larger bed- and longer time-scale, because the gaps, like territories, shifts positions with time (Bell et al. 1999, pers. obs.) as a result of simultaneous shrimp activity and seagrass recolonization. Palomar et al. (2004) have cited the importance of seagrass leaves and detritus in the shrimp's diet and included crustaceans and protozoans among the less frequently encountered food item in the guts. To strengthen this observation and resolve the shrimp's preferential consumption of these food items, follow-on choice experiments may have to be conducted based on a larger sample size of subjects representing various seagrass environments. Further, studies that detail the functional morphology of feeding appendages and the digestive tract at various stages of the life history of *A. macellarius* may elucidate the importance of various food elements utilized in the seagrass beds; their activities would also provide indications of their bioturbation potential and impact to the system, as has been revealed for other large bioturbators (Tamaki 2004).

The nonsymbiotic mutualism of *A. macellarius* and *Cryptocentrus* sp.

Mutualisms are widespread in the tropics and important to many population and community characteristics (Boucher et al. 1982, Hay et al. 2004). Nonsymbiotic mutualisms, those in which two species are physically unconnected, can involve exchanges of benefits for nutrition, supply of energy, protection, and transport. Among marine animals, the burrow-sharing by gobies and shrimps appears such a non-symbiotic mutualism, probaly strongly embedded through natural selection (Migita and Gunji 1996). About 13 species of the burrowing shrimp *Alpheus* are known to be associated with gobiid fishes (Karplus 1987) while at least 18 species of *Cryptocentrus* are associated with alpheid shrimps (Karplus 1987, FishBase 2004, Nelson 2004). It is known that shrimp gobies primarily serve as watchman (using the vicinity of the burrow opening as post), may be instrumental in displacing the openings (Karplus et al. 1974), and do not participate in burrow construction (endoecism, Atkinson and Taylor 1991). Communication between partners is mainly tactual (Preston 1978) – gobies flick their tails to signals during emergence to the surface (initially at 70 to 400 lux – Karplus et al. 1974) or retreat to the burrows while shrimps use their antennae. Retreats and subsequent hiding were seen as response to low light levels (Karplus 1987), high water motion conditions (Chapter 3), and perceived biological threats at the sediment surface. In the clear-water meadow, *C. octafasciatus* remained within a guarding radius of not more than 50 cm from the burrow opening to maintain its communication distance with *A. macellarius* while the latter continued with its aboveground activities (Chapter 3). The interactions *in situ* of *A. macellarius* and *C. singapurensis*, the shrimp's frequent partner in silty sediments (Palomar 2002), and the effects of such substrate and light conditions on the partners' communication radius and their mutualism in general, remains to be studied.

Shrimp disturbance and *Thalassia hemprichii*

Although burrowing shrimps rework a substantial amount of sediments (Chapter 3), only moderate effects were observed on the dominant seagrass *T. hemprichii*. Specifically, both early and mature stages of vegetative *T. hemprichii* displayed tolerance to short-term burial (Chapter 5). The leaf harvesting mode of *A. macellarius*, which represents single and partial defoliation events, had little impact on leaf growth (Cebrián and Duarte 1998, Chapter 5). Seedlings were able to survive partial defoliation but were sensitive to burial events (Rollon et al. 2003, Olesen et al. 2004, Chapter 5), which may be an important reason for the high mortality rates of recruits. Longer-term shoot burial induced species-specific responses, e.g., increased vertical growth, branching, and shoot densities, to cope with the disturbance (Marbá et al. 1994, Duarte et al. 1997, Berkenbusch et al. 2000). For a period of 1 year, the vegetative shoots of *T. hemprichii* were able to sustain leaf growth but did not increase shoot density by branching in exclosures without shrimp (Chapter 5). The exclusion of such disturbance was beneficial only to *Halophila ovalis*, which increased its densities in the plots, compared with other accompanying species. We can only surmise that the overall positive effect of the bioturbation on *T. hemprichii* shoots outweighs the negative impact to the sensitive seeds, since the presence of burrowing shrimps persists and is, in fact, common in thriving tropical mixed-species seagrass meadows.

Small-scale disturbance and large scale dynamics

Burrowing animals and their burrows, both in the terrestrial and aquatic realms, modify the environment in a number of forms. Meadows (1991a) outlined what was known before the 1990s of their environmental impacts on soil fertility, aeration and water infiltration, sediment chemistry, pollutant retention and release, early diagenesis, microbial activity, erosion and stability, landscapes, sediment turnover and reworking, and the development of plant and animal communities and associations. The small-scale disturbance that burrowing shrimp populations impose on the sediment compartment in seagrass meadows has a major impact on the local environment. In Chapter 2, we showed that burrowing shrimps altered the vertical profiles of sediment properties and reduced the nitrogen content of sediments – proof that the animals are of local (and global) importance. In coastal habitats, burrowing populations are assumed to prevent anoxia and hence, system collapse, primarily through (1) sediment oxygenation (ventilation), (2) the channeling of food particles to bacterial and macrobenthic production, and (3) the speed-up of mineralization and nutrient release from the sediment (de Wilde 1991, Frouin 2000). The latter may also account for enhanced mineralization in the root zone preferentially in the more often nutrient-limited Philippine seagrasses (Agawin et al. 1996). Leaf harvesting constitutes only moderate herbivory (12-42%, Chapter 3) and represents a translocation of material for microbial action and a process of nutrient conservation within the meadow (Stapel and Erftemeijer 2000). The presence of gaps in the meadows may be more important in the process of mixing within- and above-canopy waters of the beds (e.g., Granata et al. 2001), including dissolved nutrients and propagules therein, thus altering the "skimming flow" of water in homogeneous meadows (Koch et al. 2006). The sediment gaps (sand mounds) may be eroded preferentially (gap-widening) and particles be suspended and transported (Roberts et al. 1981, Suchanek, 1983, Rowden et al. 1998, Bouma et al. 2007). These gaps form a pattern of spatial habitat heterogeneity at scales of 10-100 m, increasing internal edge length in the meadow, and thus enhancing habitat availability for invertebrates and fish.

Conclusions and future directions

This thesis showed that bioturbation from burrowing shrimps is a main feature of lush seagrass beds of finer, autogenous, and carbonaceous sands. Its impact on the mineralization of nutrients may be important, as reworking helps to facilitate their availability to seagrasses. On terrigenous sediments, bioturbation, although less intense, may alleviate reducing conditions of sediments by ventilation/ irrigation events in the burrows. The burrowing shrimps may well be instrumental in maintaining a small-scale heterogeneity (i.e., relative to the shrimps) in seagrass patches and in larger, mixed-species beds, constantly creating sediment gaps that are colonized by fast pioneers, e.g., *Halophila ovalis*, and then facilitating a sequence of succession in the gaps. Established seagrasses are able to exert tolerance to burial and leaf defoliation, but to which recruits could be limited.

Alpheus macellarius was first described in a taxonomic monograph by Chace (1988), based on material from Cebu and Leyte (Philippines). Our examination of its behavior progressed from quantifying visible activities *in situ* and then justifying the subsequent outdoor observations with the considerable time the shrimps spent within their burrows. As we make generalizations of the shrimps' bioturbation, advanced research by physical engineers has begun to look at the physics of the loud sound from alpheid snapper claws (*A. heterochaelis*; Versluis et al. 2000). The snapping sound is caused by the collapse of the cavitation bubble that is formed during a pressure drop below vapor pressure, which occurs when high velocity water jet is produced by the fast claw closure following the cocking of the claw in the open position. More recently, the role of water current signals by alpheid shrimps during social interactions was described (Herberholz and Schmitz 2001). Probably new perspectives are there at hand, as technology becomes more capable of unfolding the unknowns of alpheid shrimps.

As economically valuable coastal systems in the tropics, seagrass habitats have yet to gain further due legislation and conservation measures by linking burrowing shrimps, as engineers, to ecosystem function and biodiversity (e.g., Boogert et al. 2006). Also, since bioconstruction and bioturbation are involved in most faunal community relationships in the sediment (Reise, 2002), the interactions between burrowing shrimps and biotic sympatrics may also have to be elucidated to clarify their potential key role in the meadows, considering issues of both spatial and temporal scales (*sensu* Hastings et al. 2007). Current criteria for the zoning of seagrass ecosystems in the Philippines being tested, e.g., combinations of high canopy cover on bottom, seagrass species richness, and/ or apparent significance to dugongs and turtles (PCSD 2007, Nacorda et al. 2008), show that specific core zones recommended are also frequently characterized by disturbance from burrowing shrimps. These critical areas are, thus, clear examples of disturbance-mediated and diverse habitats worthy of well-deserved protection.

References

Andersen, F.O. and Kristensen, E. 1991. Effects of burrowing macrofauna on organic matter decomposition in coastal marine sediments. In: Meadows, P.S. and Meadows, A., eds., The environmental impact of burrowing animals and animal burrows. The Proceedings of a Symposium held at the Zoological Society of London, 3-4 May 1990. Symposia of the Zoological Society of London 63: 69-88. Clarendon Press, Oxford.

Bell, S.S., Robbins, B.D., and Jensen, S.L. 1999. Gap dynamics in a seagrass landscape. Ecosystems 2: 492-504.

Berkenbusch, K., Rowden, A.A., and Myers, T.E. 2007. Interactions between seagrasses and burrowing ghost shrimps and their influence on infaunal assemblages. Journal of Experimental Marine Biology and Ecology 341: 70-84.

Berkenbusch, J., Rowden, A.A., and Probert, P.K. 2000. Temporal and spatial variation in macrofauna community composition imposed by ghost shrimp *Callianassa filholi* bioturbation. Marine Ecology Progress Series 192: 249-257.

Boogert, N.J., Paterson, D.M., and Laland, K.N. 2006. The implications of niche construction and ecosystem engineering for conservation biology. BioScience 56: 570-578.

Boucher, D.H., James, S., d Keeler, K.H. 1982. The ecology of mutualism. Annual Review of Ecology and Systematics 13: 315-347.

Bouma, T.J., van Duren, L.A., Temmerman, S., Claverie, T., Blanco-Garcia, A., Ysebaert, T., and Herman, P.M.J. 2007. Spatial flow and sedimentation patterns within patches of epibenthic structures: combining field, flume and modelling experiments. Continental Shelf Research 27: 1020-1045.

Cebrián, J. and Duarte, C.M.D. 1997. Patterns in leaf herbivory on seagrasses. Aquatic Botany 60: 67-82.

Chapin, F.S. III, Walker, B.H., Hobbs, R.J., Hooper D.U., Lawton J.H., Sala, O.E., and Tilman, D. 1997. Biotic control over the functioning of ecosystems. Science 277: 500-503.

Crutzen, P.J. and Stoermer, E.F. 2000. The "Anthropocene". IGBP Newsletter 41: 17-18.

Dorgan, K.M., Jumars, P.A., Johnson, B.D., and Boudreau, B.P. 2006. Macrofaunal burrowing: the medium is the message. In: Gibson, R.N., Atkinson, R.J.A., and Gordon,

J.D.M., eds., Oceanography and Marine Biology: An Annual Review 44: 85-121. CRC Press/ Taylor and Francis Group, Boca Raton, Florida.

Duarte, C.M. 2000. Marine biodiversity and ecosystem services: an elusive link. Journal of Experimental Marine Biology and Ecology 250: 117-131.

Duarte, C.M., Fourqurean, J.W., Krause-Jensen, D., and Olesen, B. 2006. Dynamics of seagrass stability and change. In: Larkum, A.W.D., Orth, R.J., and Duarte, C.M., eds., Seagrasses: biology, ecology and conservation, pp.271-294. Springer, Dordrecht, The Netherlands.

Duarte, C.M., Terrados, J., Agawin, N.S.R., Fortes, M.D., Bach, S., and Kenworthy, W.J. 1997. Response of a mixed Philippine seagrass meadow to experimental burial. Marine Ecology Progress Series 147: 285-294.

FishBase 2004 – a global information system on fishes. DVD, WorldFish Center – Philippine Office, Los Baños, Philippines.

Frouin, P. 2000. Effects of anthropogenic disturbances of tropical soft-bottom benthic communities. Marine Ecology Progress Series 194: 39-53.

Granata, T.C., Serra, T., Colomer, J., Casamitjana, X., Duarte, C.M., and Gacia, E. 2001. Flow and particle distributions in a nearshore seagrass meadow before and after a storm. Marine Ecology Progress Series 218: 96-106.

Griffis, R.B. and Suchanek, T.H. 1991. A model of burrow architecture and trophic modes in thalassinidean shrimp (Decapoda: Thalassinidea). Marine Ecology Progress Series 79: 171-183.

Harley, C.D.G., Hughes, A.R., Hultgren, K.M., Miner, B.G., Sorte, C.J.B., Thornber, C.S., Rodrigues, L.F., Tomanek, L. and Williams, S.L. 2006. The impacts of climate change in coastal marine systems. Ecology Letters 9: 228-241.

Hastings, A., Byers, J.E., Crooks, J.A., Cuddlington, K., Jones, C.G., Lambrinos, J.G., Talley, T.S., and Wilson, W.G. 2007. Ecosystem engineering in space and time. Ecology Letters 10: 153-164.

Hay, M.E., Parker, J.D., Burkepile, D.E., Caudill, C.C., Wilson, A.E., Hallinan, Z.P., and Chequer, A.D. 2004. Mutualisms and aquatic community structure: the enemy of my enemy is my friend. Annual Review of Ecology, Evolution, and Systematics 35: 175-197.

Hemminga, M.A. and Duarte, C.M. 2000. Seagrass ecology. Cambridge University Press, Cambridge, 298 p.

Hobbie, S.E., Jensen D.B., and Chapin, F.S. III. 1994. Resource supply and disturbance as controls over present and future plant diversity. In: Schulze, E.-D. and Mooney, H.A., eds., Biodiversity and ecosystem function, pp. 385-408. Springer-Verlag, Berlin.

Hughes, A.R. and Stachowicz, J.J. 2004. Genetic diversity enhances the resistance of a seagrass ecosystem to disturbance. Proceedings of the National Academy of Science of the United States of America 101: 8998-9002.

Hughes, L. 2000. Biological consequences of global warming: is the signal already apparent? Trends in Ecology and Evolution 15: 56-61.

Kamp-Nielsen, L., Vermaat, J.E., Wesseling, I., Borum, J., and Geertz-Hansen, O. 2002. Sediment properties along gradients of siltation in South-east Asia. Estuarine, Coastal and Shelf Science 54: 127-137.

Karplus, I. 1987. The association between gobiid fishes and burrowing alpheid shrimps. Oceanography and Marine Biology Annual Reviews 25: 507-562.

Karplus, I., Szlep, R., and Tsurnamal, M. 1974. The burrows of alpheid shrimp associated with gobiid fish in the Northern Red Sea. Marine Biology 24: 259-268.

Koch, E.W., Ackerman, J.D., Verduin, J., and van Keulen, M. 2006. Chapter 8 – Fluid dynamics in seagrass ecology – from molecules to ecosystems. In: Larkum, A.W.D., Orth, R.J., and Duarte, C.M., eds., Seagrasses: biology, ecology and conservation, pp. 193-225. Springer, Dordrecht.

Marbá, N., Cebrián, J., Enriques S., and Duarte, C.M. 1994. Migration of large-scale subaqueous bed forms measured with seagrasses (*Cymodocea nodosa*) as tracers. Limnology and Oceanography 39: 126-133.

Meadows, A. 1991a. Burrows and burrowing animals: an overview. In: Meadows, P.S. and Meadows, A., eds., The environmental impact of burrowing animals and animal burrows. The Proceedings of a Symposium held at the Zoological Society of London, 3-4 May 1990. Symposia of the Zoological Society of London 63: 1-13. Clarendon Press, Oxford.

Meadows, P.S. 1991b. The environmental impact of burrows and burrowing animals – conclusions and a model. In: Meadows, P.S. and Meadows, A., eds., The environmental impact of burrowing animals and animal burrows. The Proceedings of a Symposium held at the Zoological Society of London, 3-4 May 1990. Symposia of the Zoological Society of London 63: 327-338. Clarendon Press, Oxford.

Migita, M. and Gunji, Y.-P. 1996. Plasticity in symbiotic behavior as demonstrated by a gobiid fish (*Amblyeleotris steinitzi*) associated with alpheid shrimps. Rivisita di Biologia/ Biology Forum 89: 389-406.

Nacorda, H.M., David, L., Mendoza, W., Salamante, E., and Regino, G. 2008. Macro-scale assessment of seagrass bottoms in Bolinao (NW Philippines): indicators for the UNEP/GEF/SCS Seagrass Demonstration Site Project. (Final report)

Nelson, R. 2004. ExpolreBiodiversity.com – the definitive site for shrimp-goby information. (http://www.explorebiodiversity.com/Hawaii/Shrimp-goby/general/Taxonomy.htm)

Nickell, L.A. and Atkinson, R.J.A. 1995. Functional morphology of burrows and trophic modes of three thalassinidean shrimp species, and a new approach to classification of the thalassinidean burrow morphology. Marine Ecology Progress Series 128: 181-197.

Occhipinti-Ambrogi, A. 2007. Global change and marine communities: alien species and climate change. Marine Pollution Bulletin 55: 342-352.

Olesen, B., Marbá, N, Duarte, C.M., Savella, R.S., and Fortes, M.D. 2004. Recolonization dynamics in a mixed seagrass meadow: the role of clonal versus sexual processes. Estuaries 27: 770-780.

Orth, R.J., Caruthers, T.J.B., Dennison, W.C., Duarte, C.M., Fourqurean, J.W. , Heck, K.L. Jr., Hughes, A.R., Kendrick, G.A., Kenworthy, W.J., Olyarnik, S., Short, F.T., Waycott, M., and Williams, S.L. 2006. A global crisis for seagrass ecosystems. BioScience 56: 987-996.

Palawan Council for Sustainable Development (PCSD). 2007. Coastal and marine resources of North Palawan. PCSD, Puerto Princesa City, Palawan, Philippines.

Palomar, N.E., Juinio-Meñez, M.A., and Karplus, I. 2004. Feeding habits of the burrowing shrimp *Alpheus macellarius*. Journal of the Marine Biological Association of the United Kingdom 84: 1199-1202.

Philippart, C.J.M. 1994. Interactions between *Arenicola marina* and *Zostera noltii* on a tidal flat in the Dutch Wadden Sea. Marine Ecology Progress Series 111: 251-257.

Picket, S.T.A. and White, P.S., eds. 1985. The ecology of natural disturbance and patch dynamics. Academic Press, Inc., Orlando, Florida, 455 p.

Pillay, D., Branch, G.M., and Forbes, A.T. 2007. Experimental evidence for the effects of the thalassinidean sandprawn *Callianassa kraussi* on macrobenthic communities. Marine Biology. (DOI 10.1007/s00227-007-0715-z)

Preston, J.L. 1978. Communication systems and social interactions in a goby-shrimp symbiosis. Animal Behaviour 26: 791-802.

Reise, K. 2002. Sediment mediated species interactions in coastal waters. Journal of Sea Research 48: 127-141.

Rollon, R.N. 1998. Chapter 2: Characterization of the environmental conditions at the selected study sites: seagrass habitats of Bolinao, NW Philippines. In: Rollon, R.N., Spatial variation and seasonality in growth and reproduction of *Enhalus acoroides* (L.f.) Royle populations in the coastal waters off Cape Bolinao, NW Philippines, pp. 9-36. PhD Thesis, Wageningen Agricultural University and IHE-Delft. A.A. Balkema, Rotterdam.

Rollon, R.N., Vermaat, J.E., and Nacorda, H.M.E. 2003. Sexual reproduction in SE Asian seagrasses: the absence of a seed bank in *Thalassia hemprichii*. Aquatic Botany 75:181-185.

Schulze, E.-D. and Mooney, H.A., eds. 1994. Biodiversity and ecosystem function. Springer-Verlag, Berlin, 525 p.

Short, F.T. and Neckles, H.A. 1999. The effects of global climate change on seagrasses. Aquatic Botany 63: 169-196.

Siebert, T. and Branch, G.M. 2006. Ecosystem engineers: interactions between eelgrass *Zostera capensis* and the sandprawn *Callianassa kraussi* and their indirect effects on the mudprawn *Upogebia africana*. Journal of experimental Marine Biology and Ecology 338: 253-270.

Smith, C.R., Austen, M.C., Boucher, G., Heip, C., Hutchings, P.A., King, G.M., Koike, I., Lambshead, J.D., and Snelgrove, P. 2000. Global change and biodiversity linkages across the sediment-water interface. BioScience 50: 1108-1120.

Snelgrove, P.V.R. 1999. Getting to the bottom of marine biodiversity: sedimentary habitats. BioScience 49: 129-138.

Stapel, J.A. and Erftemeijer, P.L.A. 2000. Leaf harvesting by burrowing alpheid shrimps in a *Thalassia hemprichii* meadow in South Sulawesi, Indonesia. Biologia Marina Mediterranea 7: 282-286.

Suchanek, T.H. 1983. Control of seagrass communities and sediment distribution by *Callianassa* (Crustacea, Thalassinidea) bioturbation. Journal of Marine Research 41: 281-298.

Tamaki, A., ed. 2004. Proceedings of the Symposium on Ecology of Large Bioturbators in Tidal Flats and Shallow Sublittoral Sediments – from Individual Behavior to their Role as Ecosystem Engineers. Nagasaki University, Nagasaki, Japan, 118 p.

Terrados, J., Duarte, C.M., Fortes, M.D., Borum, J., Agawin, N.S.R., Bach, S., Thampanya, U., Kamp-Nielsen, L., Kenworthy, W.J., Geertz-Hansen, O., and Vermaat, J. 1998. Changes in community structure and biomass along gradients of siltation in SE Asia. Estuarine, Coastal and Shelf Science 46: 757-768.

Townsend E.C. and Fonseca, M.S. 1998. Bioturbation as a potential mechanisms influencing spatial heterogeneity of North Carolina seagrass beds. Marine Ecology Progress Series 169: 123-132.

Vander Wall, S.B. 1990. Food hoarding in animals. The University of Chicago Press, Chicago, 445 p.

Versluis, M., Schmitz, B., von der Heydt, A., and Lohse, D. 2000. How snapping shrimps snap: through cavitating bubbles. Science 289: 2114-2117.

Wilde, P.A.W.J. de. 1991. Interactions in burrowing communities and their effects on the structure of marine benthic ecosystems. In: Meadows, P.S. and Meadows, A., eds., The environmental impact of burrowing animals and animal burrows. The Proceedings of a Symposium held at the Zoological Society of London, 3-4 May 1990. Symposia of the Zoological Society of London 63: 107-117. Clarendon Press, Oxford.

Woods, C.M.C. and Schiel, D.R. 1997. Use of seagrass *Zostera novazelandica* (Setchell, 1933) as habitat and food by the crab *Macrophthalmus hirtipes* (Heller, 1862) (Brachyura, Ocypodidae) on rocky intertidal platforms in southern New Zealand. Journal of Experimental Marine Biology and Ecology 214: 49-65.

Samenvatting

Bodembewonende garnalen en zeegrasdynamiek in ondiepe kustwateren rond Bolinao (NW Filippijnen)

Natuurlijke verstoringen dragen bij aan de dynamiek van zeegrasvelden. Het doel van dit promotie-onderzoek was het effect te kwantificeren van kleinschalige verstoringen door algemeen in Filippijnse zeegrasvelden voorkomende pistoolgarnalen (*Alpheus macellarius*) en spookgarnalen [1] (Thalassinidea: Callianassidae). De blinde pistoolgarnalen leven paarsgewijs in zelfgegraven tunnelstelsels in symbiose met een grondeltje (*Cryptocentrus spec.*). Ze knippen zeegrasbladeren af en slaan deze op in ondergrondse voorraadkamers. Zowel het intensieve grondverzet (bioturbatie) als het oogsten van bladmateriaal zou van belang kunnen zijn voor het omringende zeegrasveld. Spookgarnalen graven burchten bestaande uit meerdere U-vormige gangen, met één of meerdere vulkaanachtige uitmondingen. Ze leven van organisch materiaal in het sediment door dat te zeven met hun monddelen of filtreren gesuspendeerd materiaal uit het water wat door de burcht wordt opgepompt. Uitgegraven of anderszins verwerkt sediment wordt naar het bodemoppervlak geloosd. Dit proefschrift behandelt achtereenvolgens (a) de ruimtelijke verspreiding van deze twee typen gravende garnalen in zeegrasvelden langs een sedimentatiegradient bij Bolinao (Luzon, NW Filippijnen; hoofdstuk 2); (b) het boven- en ondergrondse gedrag van pistoolgarnalen (hoofdstuk 3 en 4); en (c) de gevolgen van grondverzet en bladoogst door pistoolgarnalen op de groei en uitbreiding van verschillende soorten zeegras.

Beide groepen bodembewonende garnalen komen algemeen voor in de zeegrasvelden rond Bolinao en elders op de Filippijnen. Met name in beschutte velden op fijn, zandig sediment zijn de dichtheden hoog. Bioturbatie door *Alpheus* resulteerde in meer open plekken in het zeegrasveld (15% ± 1% van het oppervlak, 102 ± 5 plekken per 100 m^2) dan die door *Callianassa* (5% ± 1%, 52 ± 7 plekken per 100 m^2). Beide garnalen-groepen beïnvloeden door hun graafwerk de verticale verdeling van de korrelgrootte, organische stof en stikstof. *Alpheus* stapelt grof materiaal aan het oppervlak, als een soort tunneldak, terwijl *Callianassa* fijn bodemmateriaal concentreert tussen de 10 en 20 cm diepte. Beide groepen halveerden de stikstofconcentraties in de bovenste 10 cm van de bodem.

Alpheus besteedde slechts 12% van de tijd bovengronds buiten het gangenstelsel. Dan werd zeegrasblad geoogst of het gangenstelsel gerepareerd. In de natte moessontijd was dat nog minder (8%). De garnaaltjes verwerkten gemiddeld 1.1 kg m^{-2} d^{-1} sediment en oogsten 11.4 g blad m^{-2} d^{-1}. Dit laatste komt neer op 27% van de dagelijkse zeegrasproductie, een aanzienlijke herverdeling van plantaardige productie naar de bodem. In kleine cuvetten (45 x 2 x 35 cm, L x B x H) met koraalzand en zeegras in het lab begon *Alpheus* direct te graven. Binnen twee uur was de primaire tunnel afgerond en konden ze zich verbergen. Vervolgens ging het graafwerk door, afgewisseld door poets-, verkennings- en forageergedrag (zowel suspensiefilteren, zeegras oogsten als sedimentopname). Overdag werd minder tijd

---

[1] Van *Callianassa* is ons geen Nederlandse naam bekend, zie bijvoorbeeld www.kreeftengarnalen.nl of www.imv.uit.no/crustikon. We hebben spookgarnaal letterlijk uit het Engelse ghost shrimp vertaald.

bovengronds besteed dan's nachts. In behandelingen met fijn terrigeen slib werd veel minder gegraven dan in koraalzand; de dieren verstopten zich voornamelijk in korte tunneltjes. Dit bevestigt de in het veld waargenomen habitatvoorkeur van deze pistoolgarnaalsoort.

Uit de experimenten die het effect van het garnaalgedrag op zeegras (*Thalassia hemprichii*, in deze regio meestal een dominante soort) bleek dat de planten weinig hinder hadden van de kleinschalige verstoringen. Apicale scheuten reageerden op veertien dagen durende begraving met versnelde blad- en stengelgroei. Afknippen van de bladschijven of een combinatie van begraven en knippen hadden slechts minimale gevolgen op de blad- en wortelstokgroei. Kiemplanten van *Thalassia* overleefden experimentele ontbladering maar bleken erg gevoelig voor begraving: de groei verminderde aanvankelijk tot slechts een derde van de controlegroep. Tenslotte bleek uit een dertien maanden durend *in situ* experiment waarbij pistoolgarnalen uitgesloten werden dat daar geen meetbaar effect was op de dichtheid en groei van de meeste aanwezige soorten zeegras. Slechts één soort, de kleine, snel groeiende maar kort levende *Halophila ovalis* nam gedurende enige tijd sterk toe in dichtheid, met een maximum na vijf maanden.

Resumerend kan geconcludeerd worden dat de twee algemeenste groepen in de bodem gravende garnalen van de tropische zeegrasvelden in Zuid-Oost Azië met name grote hoeveelheden sediment verplaatsen. Daarmee beïnvloeden ze ook de verticale verdeling in het sediment van korrelgrootte, organische stof en stikstof. Indirect zorgen ze er ook voor dat ongeveer een derde deel van de nutrienten dat in het bladmateriaal vastgelegd is niet door afslag uit het zeegrasveld verdwijnt, maar via mineralisatie in het sediment weer beschikbaar komt voor het zeegras. De pistoolgarnalen zijn het sterkst gebonden aan het zeegras omdat ze de bladeren oogsten. Zowel het graafwerk als de herbivorie hebben echter nauwelijks effect op de groei en overleving van de uitgebreide klonale rhizoomnetwerken van het zeegras. Wel wordt waarschijnlijk de recrutering en groei van kiemplanten negatief beïnvloed. Dit wordt vermoedelijk gecompenseerd door een effectieve vegetatieve vermeerdering.

Buód

Mga naghuhukay na hipon at dinámika ng mga lináng ng damong-dagat sa Bolinao (H-K Pilipinas)

May kontribusyón ang mga pangyayari sa kalikasan sa dinámika ng mga lináng ng damong-dagat. Layunin ng tesis na ito na tasahín ang halagá ng máliitang pagbabago (*bioturbation*) na likhâ ng naghuhukay na hipon, bilang isang tagatakdâ sa mobilidád ng sediment at isang sálik sa pagkakaroón, pagláwak, at komposisyón ng *species* sa mga lináng ng damong-dagat sa Pilipinas. Ang mga espesípikong layunin nitong pag-aaral ay mailarawan ang *spatial* na distribusyón ng pagbabagong likhâ ng mga naghuhukay na hipon sa isang *siltation gradient* at ang bunga nito sa mga vertikal na katangian ng latag ng *sediment*, ang masukat ang gawì ng naghuhukay na mga hipong *alpheid*, at mabatíd ang mga epekto ng panandaliang pagtatabon at pag-ani sa dahon sa padron ng pagtubò ng dominanteng damong-dagat na *Thalassia hemprichii* Ehrenberg (Ascherson).

Ang mga bukana ng hukay at mga patláng ng *sediment* ay karaniwan sa mga lináng at iniúugnay sa hipong *caridean* na *Alpheus macellarius*, Chace, 1988, o sa mas malakí at mas malalim maghukáy na *species* ng *Thalassinidea*. Sa mga lináng na ligtás-sa-alon ay mas madalás at malakí ang mga patláng hambíng sa mga lantád na lináng; mas laganap din ang hukay at tumpók ng *sediment* na likhâ ng hipon sa mga lináng na may malinaw kaysa malabong tubig. Ang mga pagbabagong likhâ ng *A. macellarius* ay natuklasáng mas higít kaysa likhâ ng mga hipong *thalassinidean*, sa kalahatán, at ang distribusyón ng mga hipong *alpheid* ay kadalasang limitado lamang sa latag na may halaman at pino ang buhangin. Ang latag ng buhangin na kaugnáy at binago ng *A. macellarius* ay sumakop sa 15 ± 2% ng mga lináng (sa eskaladong dami na 102 ± 5 bawat 100 m^2, *cf.* larawan sa pabalat) samantalang ang tumpók ng buhangin ng hipong *thalassinidean* ay matatagpuán sa 5 ± 1% (sa dami na 52 ± 7 bawat 100 m^2). Binago ng mga naghuhukay na hipon sa magkakaibáng paraán ang mga vertikál na katangian ng sediment—inilipat ng *A. macellarius* ang malakíng bahagi ng malalakíng butil sa 10 cm rabaw, samantalang tinipon ng hipong *thalassinidean* ang mga pinong butil mula 10 pababa hanggang 20 cm ng *sediment*, at nagtumpók ito ng organikong materyal. Gayunmán, ang dalawáng uri ng hipon ay nagpababa sa nitrohéno ng sediment ng 20-73% sa rabaw na saray at ng 4-46% sa lalim na higít sa 10 cm.

Base sa aktuwal na obserbasyon ng nasa rabaw-na-lupang gawì ng *A. macellarius*, ang mga hipon ay nagpakita ng aktibong pagrebasé ng *sediment* at okasyonál na pag-ani ng dahon ng damong-dagat sa lináng na malinaw ang tubig sa panahon ng tag-inít. Ang ginugol ng mga hipon para sa ganitong mga aktibidad ay 12% lamang ng kanilang aktibong yugtô at sila ay nasa kani-kanilang lunggâ sa butál ng bawat araw. Sa panahon ng tag-ulán, nabawasan ang nasa rabaw-na-lupang gawì nang hindi bababâ sa 34% at ang panahon nila sa loob-ng-lunggâ ay tumaas ng 5% bilang bunga nito. Sa karaniwan, kaya ng *A. macellarius* magrebasé ng ~300 g *DW* (tuyóng timbang) ng *sediment* bawat araw (o 112 kg bawat taon), at umani ng 0.8 g *DW* ng dahon bawat araw (o 291.3 g bawat taon). Mahalaga at dapat isaalang-alang ang kinalkulang bilís ng pagrebasé sa *sediment* (0.8 hanggang 1.4 kg kada 1 m^2 kada araw) para sa karaniwang dami ng hipon na 2 indibidwal kada 1 m^2, at ang bilis ng pag-ani nito ng dahon ay kumakatawan naman sa katamtamang *herbivory*

(0.4 hanggang 2.3 g kada 1 m² kada araw), katumbas ng 12 hanggang 42% ng produksiyón ng dahon.

Ang mga sumunód na obserbasyón sa laboratoryo na sumurì sa papél ng urì ng *sediment*—buhangin, maputik na buhangin, at mabuhanging putik—sa gawì ng *A. macellarius* ay nagpakita na nagsimulâ agad ang mga hipon sa paghuhukáy sa oras na magkaroón ito ng kontak sa mga *substrate*. Sa substrate na buhangin, naganáp nang maaga ang pagtatagò ng hipon (may unang lunggâ na sa loob ng 2 oras) sa kabilâ ng masikap nilang paghuhukáy, at nagawâ rin ang ekstensibong habà ng lunggà. Nabawasan ang paghuhukáy ng hipon at malinaw na nakita ang paglaboy nito pagkaraan ng ikalimáng linggo ng obserbasyón. Litáw na litáw na rin ang kanyang panginginain—pangunahin ang pagkain ng partikél, paminsan-minsang pagkaing-suspensiyón, at pagngatngat sa dahon ng damong-dagat. Ang gawì na paghuhukay, paglilinis ng sarili, at pagsisiyasat ay kapansín-pansín sa araw samantalang ang paglaboy at panginginain ay tumátagal hanggang gabi. Gayunman, ang paghuhukáy ay nanatiling mas malimit sa buhangin kaysa dalawa pang uri ng substrate. Sa kalahatán, ang malambot at *carbonate* na buhanging sediment, kasama ng pagbilaan ng makapal na mga damong-dagat, ay may mas mataas na suporta sa gawì na paghuhukay ng *A. macellarius* at kaugnay ito ng naobserbahang mas mataas na dami nila sa mga habitat na ito sa aktuwal na kaligiran. Hambíng dito, lumitaw na limitado ang paghuhukay ng hipon sa *terrigenous* (putik) na *substrate* at nahalilihan ito ng estratehiya ng pagtatagò; dahil sa mas kauntíng suporta ng bibihirang halaman, mas mababa ang bilang ng lunggâ sa mga habitat na ito sa aktuwal na kaligiran.

Ang serye ng manipulatibong eksperimento pangunahin sa damong-dagat na *T. hemprichii* ay nagbigay ng patunay sa tolerance o hindi pag-indâ ng mga usbóng ng halaman sa maliitang pagbabagong likhâ ng naghuhukay na hipon. Ang minsanang pagkatabon sa tagal na di-bababa sa 14 araw ay nanghikayat ng mabilis na pagtubò ng dahon sa mga usbóng na nasa *apex*, samantalang ang pagtatabás sa dahon at kombinasyon ng mga *treatment* sa ekspetimento ay may *minimum* na epekto sa pagtubò ng dahon o ng suwi. Nalampasan din ng mga binhi ang paglalagás sa dahon pero sensitibo ang mga ito sa pagkatabon—hindi lumikha ng pagbabago sa pagtubò ng mga binhi ang pagtatabas lamang, ngunit ang pagtubò ay patuloy na bumaba dahilan sa pagkatabon, bukod man ito o kasabáy sa pagtatabás. Hindi nakaimpluwensiya ang eksklusyón ng mga aktibidád ng hipon sa bilís ng pagtubò ng dahon ng mga magulang na usbóng. Tanging ang dami ng *Halophila* lang ang napagyaman, partikular pagkaraan ng 21 linggo sa loob ng 13-buwang tagál ng eksperimento, at kapwa ang pagtubò ng *T. hemprichii* at dami ng usbóng ng ibang kasamang *species* ng damong-dagat ay nagpakita ng malakás na pagbabagu-bagong temporal tulad ng inaasahan.

Sa madaling sabi, naipakita ng tesis na ito na ang dalawáng uri ng naghuhukay na hipon—ang *alpheid* at ang *thalassinidean*—na karaniwan sa mga lináng ng damong-dagat, ay may kakayahang maghalubilo ng maraming sediment at magdulot ng mahalagang epekto sa depth gradient ng organikong materyal, laki ng butil, at nitroheno. Ang mga hipong *alpheid* lamang ang may mahigpit na kaugnayan sa damong-dagat, katulad ng panginginain nila dito at pag-aalis ng katamtamang bahgi ng produksiyóng primarya, ngunit hindi nito markadong naaapektuhan ang mga establisado at *clonal* na mga lináng ng damong-dagat. Malamang na naaapektuhan ng mga ito ang pagngalap ng mga buto at binhi sa negatibong paraan, subalit hindi lubhang naaapektuhan ang mga establisadong kaligiran na tinatahanan ng mga hipong ito.

Kabuowan

Mga orang nin ampamotbot at dinamika sa bakas ran tarektek sa Bolinao (H-K Pilipinas)

Main kontribusyon a pangyayadi sa paligid tamo sa dinamika sa bakas ran rabot sa baybay o tarektek. Say rabay nan mangyadi nin sitin tesis ket keketen nay importansya nin daedaekleng ni paninili ni gawa nin orang nin ampamotbot, bilang saya nin tagatakda sa mobilidad nan sedimento tan saya asalik sa pagkamain, pagpapaalawang, tan say abaw nin bakas sa linang nin rabot sa taaw tarektek sa Pilipinas. Say espesipikong importansiya nan sitin pagaadal ket pigaw maipakit a *spatial* nin distribusyon nin pamabayo nin bakas nin saray orang nin ampamotbot sa sayan *siltation gradient* tan say epekto nan siti sa vertical nin kabistawan sa kama nan sedimento, tan masukat a ugali nan ampangubot nin orang a alpheid, tan para matandaan a epekto nin anted nin panabon tan pangalap sa bulong sa pardon nin pamatubo nin dominanten rabot sa baybay *Thalassia hemprichii* Ehrenberg (Ascherson).

Say adapan obot tan patiang nan sedimento ket kasabyan sa linang tan manikoneksyon ra sa orang *caridean* nin *Alpheus macellarius*, Chace, 1988, o sa mas alaki tan mas ararem mamutbot a klase nan *Thalassinidea*, itaw sa linang nin protektado sa daluyon ket mas pirmi nin alaki a patiang kompara sa konran nakaliwa nin linang; mas pirmi et a ubot tan tumpok nin sedimento nin gawa nan orang sa linang nin malinaw kompara sa malabo nin ranom. Sa kabayuwan gawa nan *A. macellarius* ket nadiskubre ran bas abaw diyan sa gawa nan orang nin *thalassinidean*, sa kabuowan, tan say distribusyon ran orang *alpheid* ket karamiwan limitado tamo sa latag nan buyangin nin kaugnay tan sinilyan nan *A. macellarius* ket sumakop na 15 ± 2% nan linang (sa eskaldon kabawan na 102 ± 5 sa barang 100 m^2, *cf.* retrato sa pabalat) bale say tumpok nan buyangin nin orang *thalassinidean* ket makit sara sa 5 ± 1% (sa kabawan nin 52 ± 7 bawat 100 m^2). Sinilyan ran orang nin ampangubot sa ambo pareho nin praan a vertikal; nin katangian nan sedimento – nialis nan *A. macellarius* a alakin parte nin alalakin butil sa 10 cm rabaw, bale si orang a *thalassinidean* ket pinikakalamo nay pinon butil mangibwat sa 10 pakayupa anggan 20 cm nan sedimento, tan nangtumpok yaet nin organiko nin material. Wanin man, saray ruwa nin klase nan orang ket ampamakayupa saran in nitrogeno nan sedimento ni 20-73% sa babo nin saray tan 4-46% sa rarem nin sobra sa 10 cm.

Base sa aktuwal nin obserbasyon nin iti sa babo gawa nan *A. macellarius*, saray orang ket ampamakit saran in aktibong pagrebase nan sedimento tan matalag nin pangalap nin bulong nin tarektek sa linang nin malinaw a ranom sa tiyempon mamot. Sa gawa ran orang saw anti nin aktibidades ket 12% tamo a aktibon odas tan sara ket itaw sara sa ubot ra sa butal nin bawat awro. Sa tiyempon rapeg, mankaglatan a iti sa babo nin gawa ra nin kai kumayupa sa 34% tan say odas ra sa rarem ubot ra ket umatagay nin 5% bilang epekto nn siti. Kasabyan, kaya nan *A. macellarius* magrebase nin ~300 g *DW* (tuyo nin timbang) nin sedimento bawat awro (o 112 kg bawat taon), tan mangalap nin 0.8 g *DW* nin bulong bawat awro (o 291.3 g bawat taon). Importante tan dapat biyan alang-alang a kalkuladong kapaspasan nin pagrebase sa sedimento (0.8 anggan 1.5 kg bawat 1 m^2 bawat awro), tan say kapaspasan nin pangalap nan bulong ket nankumatawan namaet sa katamtaman *herbivory* (0.4 anggan 2.3 g bawat 1 m^2 bawat awro), katumbas nan 12 anggan 42% nan produksyon nin bulong.

Saraytin obserbasyon sa laboratoryo nin nag-adal sa papel nin klase nin sedimento – buyangin, matupa nin buyangin, tan mabuyangin nin tupa – sa gawa nan *A. macellarius* ket man-ipaket nan timmarana tampor a pamumutbot ran orang sa odas nin magkamain saran in kontak sa *substrate*. Sa substrate nin buyangin, nangyadi nin masakbay a pag-adi nan orang (main unan ubot na sa loob nin ruway odas) sa kabila nin maseseg ran pang-ubot, tan nagawa ran anron ubot. Naglatan a pangubot ran orang tan malinaw nin nakit a pikadaw ra mayadi a ikaliman linggo nan obserbasyon. Makit nin maong a pangangan ra, muna a pangan nin partikel, matalag a pngan ran in suspensyon, tan sa pangangatngat ran bulong nan tarektek o rabot baybay. Say gawa ran in pangungubot, panglinis ran in lalaman ra, tan pag ikot-ikot ra ket madlaw sa awro bale say pamamasyar tan pangan ra ket umteng anggan yabi. Bale, say pangungubot ra ket mas pirmi sa mabuyangin diyan sa ruwa et nin sakalakon *substrate*. Sa kabuowan, say malumo tan ma-*carbonate* nin buyangin sedimento, kalamo sa pagbilaan nin makubpal nin tarektek, ket main mas atagay nin suporta sa gawa nan pangungubot nin *A. macellarius* tan wanin et a naobserbawan nin mas atagay nin kabawan ra sa lugar nin wanti sa aktuwal nin kaligiran.

Say serye nin manipulatibong eksperimento nauna sa rabot baybay nin *T. hemprichii* ket nami nin patunay sakasaw o kai pag-inda ran usbong nan tanaman sa daekleng nin paninili nan ampangubot a orang. Say minsanan nin kasrep sa ubot sa eteng nin kai kumayupa sa 14 awro ket makapangikayat nin paspas a pagtubo nin bulong sa usbong nin iti sa *apex*, wanin man say pamumutol sa bulong tan say kombinasyon nin man-gaw en sa eksperimento ket main minimum a epekto sa katubo nin bulong o sa suwi na. Nalampasan ran lamang nin binhi a manabo nin bulong bale sensitibo sarayti sa pagkatabon ra sa ubot – kai sara ginumwa nin sakalako sa pagtubo ran binhi, no kai pamumutol tamo, bale say katubo ket tuloy et a kumayupa bana sa pagkatabon, bukod et yayti o karngan sa pamumutol. Kai makaimpluwensya a eksklusyon nin saray aktibidades nan orang sa kapaspasan nin katubo nin bulong ran matua nin usbong. Siti tamo si kabawan nan *Halophila* a nagpayaman, particular mayadi a 21 linggo sa loob nan 13 bulan eteng nn eksperimento, tan saray kapwa katubo ran *T. hemprichii* tan kabawan nin usbong nin raruma nin *species* nan rabot-baybay o tarektek ket namakit nin maksaw nin pisili-sili nin *temporal* tulad nin man-asahan.

Sa adanin irgo, naipakit nn tesis nin siti a say ruwa nin klase nin ampangubot a orang – say *alpheid* tan *thalassinidean* – nin pirmin makit sa linang ran tarektek, ket magwa ran magkakakalamo nin abaw sedimento tan mami nin importanten epekto sa *depth gradient* nan organikong material, kalakian nan butil, tan nitrogeno. Bale, main mahigpit a kaugnayan ran orang *alpheid* sa tarektek, bilang nin pangangan ra sa sayti tan say panglat nin 12 anggan 42% nin primarya produksyon, kai ra markadon mankaapektuwan a establisado tan *clonal* nin linang nan rabot baybay o tarektek. Sa posiblen maapektuwan sarayti ket say pangalap nin bikoy tan binhi sa negatibo nin pamamaraan, bale magwa nin kasa nin grabe o duka nin epekto sa establisadon kaligiran nan pagbaliwan nin saraytin orang.

About the Author

Hildie Maria Estacio Nacorda was born on 20 April 1966 in Lucena City (Quezon), The Philippines. She finished secondary school in 1983 at the Quezon National High School and then pursued a degree in Fisheries (major in Inland Fisheries) at the University of the Philippines in the Visayas (UPV). In 1987, she worked as research assistant at the UPV College of Fisheries–Institute of Fisheries Development and Research (IFDR) and then as research associate at the Marine Science Institute (MSI) from 1988 to 1996 for a number of soft bottom community studies led by Prof. Helen Yap. Her M.Sc. in Marine Biology was granted by MSI–University of the Philippines in Diliman in 1996, with a thesis titled *"Macroinfaunal communities of a tropical sandy reef flat"*, which was based on research work carried out under the framework of the ASEAN-Australia Project on *"Living Coastal Resources"*. In 1997, WOTRO (Netherlands Foundation for the Advancement of Tropical Research) awarded her a fellowship grant to pursue PhD research under Dr. ir. Jan Vermaat on the project *"Effects of burrowing crustaceans on sediment and seagrass dynamics along a siltation gradient"* (WB84-413), which is a sandwich scheme in Environmental Science and Technology at the UNESCO-IHE. This Ph.D. thesis integrates several studies that assessed the impact of burrowing shrimps on seagrass dynamics in shallow-water meadows.

PUBLICATIONS

Vermaat, J.E., Rollon, R.N., Lacap, C.D., Billot, C., Alberto, F., Nacorda, H.M.E., Wiegman, F., and Terrados, J. 2004. Meadow fragmentation and reproductive output of the SE Asian seagrass *Enhalus acoroides*. Journal of Sea Research 52(4): 321-328.

Rollon, R.N., Vermaat, J.E., and Nacorda, H.M.E. 2003. Sexual reproduction in SE Asian seagrasses: the absence of a seedbank in *Thalassia hemprichii*. Aquatic Botany 75: 181-185.

Lacap, C.D.A., Vermaat, J.E., Rollon, R.N., and Nacorda, H.M. 2002. Propagule dispersal of the SE Asian seagrasses *Enhalus acoroides* and *Thalassis hemprichii*. Marine Ecology Progress Series 235: 75-80.

Lacap, C.D.A., Vermaat, J.E., Rollon, R.N., Nacorda, H.M.E., and Fortes, M.D. 2000. Implications of the short seed dispersal in the seagrass *Enhalus acoroides* (L.f.) Royle. Biologia Marina Mediterranea 7: 83-86.

Yap, H.T., Nacorda, H.M.E., and Jacinto, G.S. 1998. Community structure and distribution of soft bottom fauna at various distances from the geothermal discharge sites in the eastern Philippines. Asian Journal of Tropical Biology 2: 9-21.

Nacorda, H.M.E. and Yap, H.T. 1997. Structure and temporal dynamics of macroinfaunal communities of a sandy reef flat in the NW Philippines. Hydrobiologia 353: 91-106.

Nacorda, H.M.E. and Yap, H.T. 1996. Macroinfaunal biomass and energy flow in a shallow reef flat of the NW Philippines. Hydrobiologia 341: 37-49.

Nacorda, H.M.E. and Yap, H.T. 1994. Sediment community metabolism and macroinfaunal structure in a tropical sandy reef flat. In: Sudara, S., Wilkinson, C.R., and Chou, L.M., eds., Proceedings of the 3rd ASEAN-Australia Symposium on Living Coastal Resources, Bangkok, Thailand, 16-20 May 1994, Vol. 2, pp. 197-202.

Nacorda, H.M.E., Yap, H.T., Alvarez, R.M., Oñate-Pacalioga, J., and Estacion, J. 1994. Variations in the structure of sediment communities in selected Philippine sites. In: Sudara, S., Wilkinson, C.R., and Chou, L.M., eds., Proceedings of the 3rd ASEAN-Australia Symposium on Living Coastal Resources, Bangkok, Thailand, 16-20 May 1994, Vol. 1, pp. 347-364.

Chou, L.M., Paphavasit, N., Kastoro, W.W., Nacorda, H.M.E., Othman, B.H.R., Loo, M.G.K., and Soedibjo, B.S. 1994. Soft bottom macrobenthic communities of the ASEAN region and the influence of associated marine ecosystems. In: Sudara, S., Wilkinson, C.R., and Chou, L.M., eds., Proceedings of the 3rd ASEAN-Australia Symposium on Living Coastal Resources, Bangkok, Thailand, 16-20 May 1994, Vol. 1, pp. 325-332.

Yap, H.T. and Nacorda, H.M.E. 1993. Some aspects of the ecology of sediment fauna in Balingasay, Bolinao, Pangasinan (N Philippines). In: Morton, B., ed., Proceedings of the 1st Conference on the Marine Biology of Hong Kong and the South China Sea, Hong Kong, 28 October–2 November 1990, Vol 2, pp. 509-519.

Nacorda, H.M.E. and Yap, H.T. 1992. Preliminary overview of structure and distribution of sediment communities in Southeast Asia. In: Chou, L.M. and Wilkinson, C.R., eds., 3rd ASEAN Science and Technology Week Conference Proceedings, Vol. 6, Marine Science: Living Coastal Resources, Singapore, 21-23 September 1992, pp. 171-174.

Yap, H.T., Viloria, B.H., and Nacorda, H.M.E. 1991. Comparison of soft bottom community profiles in two Philippine nearshore sites. In: Alcala, A.C., ed., Proceedings of the 1st Reginoal Symposium in Coastal Living Resources, Manila, 30 January–1 February 1989, pp. 261-277.

T - #0116 - 071024 - C2 - 254/178/6 - PB - 9780415484022 - Gloss Lamination